PERSEPHONE IS TRANSPLUTO

*THE SCIENTIFIC, MYTHOLOGICAL, AND
ASTROLOGICAL DISCOVERY OF THE
PLANET BEYOND PLUTO*

by
Valerie Vaughan

One Reed Publications
Amherst Massachusetts

Dedication

to Sean and Gabriel

Acknowledgments

I wish to express my thanks to the following people:

Bradley V. Clark, who collaborated with me for seven years in the creation and development of many concepts related in this book. Brad devoted countless hours of telephone conversation and reams of written correspondence as support, feedback, input, and outrageous humor. Although I have tried to acknowledge specific phrases and ideas of Brad's wherever possible, there are whole concepts originated by him which I have developed and re-worded, but which are so integrated into the background of the text that it is impossible to isolate all of them as footnote-able. With his Pisces Ascendant, Brad has been, in a sense, a ghost writer: invisible and selfless in his contribution. I thank him with all my heart.

I also express deep gratitude to my publisher Bruce Scofield who constantly encouraged me to overcome the limitations of single motherhood in order to produce this book. Without his support, it just wouldn't have happened.

Portions of this book are revised versions of some of my articles previously published in astrological magazines. These have been identified in footnotes.

ASTROLABE deserves thanks for chart calculations and printing from their Printwheels software, and also for the Persephone ephemeris generated from their Solar Fire software package.

Larry Ely also deserves credit for being the first to inspire my interest in this planet back in 1977.

Last, but not least, I give my regards to those individuals (who will remain nameless) that have rejected my interpretations or who deny the existence of Persephone — they forced me to strengthen my arguments. *Absit invidia.*

Chapter Contents, Tables and Charts

Foreword

1. **Astronomy and Astrology: Real Science vs. Rejected Knowledge 10**
   Planet X - Transpluto 12
   A Name for Planet X - Transpluto 16
   Discovery and Naming of Uranus, Neptune, Pluto 18

2. **Discovery of a New Planet 21**
   Planetary Changes in Consciousness 22
   Traditional, Dual Rulerships and Modern, Outer Planet Rulerships 24
   Table of Traditional and Modern Rulerships 25
   Virgo Ruled by Chiron or the Asteroids 26
   Twelve Planets - Twelve Signs 28

3. **Rebirth of the Feminine: The Return of the Goddess 29**
   Mythic Themes in Today's World 30
   Separation-Reunion, Crossing Borders, Immortality, Ecology 31
   Gender Issues, Abductions, The "Homeless" 33

4. **Mythology as Mystery — Or, The Mis-Story of Herstory 36**
   Revisionist Mythology, the Disney Version 38
   Victim Consciousness and the Rape of Persephone 39
   Female Venus-Envy 41
   An Alternative View of Aphrodite 42

5. **Persephone's Rulership of Libra: The Union of Dualities 44**
   Separation Polarity of Mars-Venus 45
   Balanced Partnership of Pluto-Persephone 46
   Complete Table of Planetary Rulerships 47
   Ladder of the Planets 50
   The Higher Octave of Venus 51
   Sexuality and Maternity 53
   Mother-Daughter and Son-Lover 55

6. New Planet on the Horizon 58
The Masculine-Feminine Split 59
Crisis of Separation-Reunion 61
Conscious Relationship 63
Ascendant-Descendant Axis 65
Mirror Reflections 69

7. The Mythology of Scientific Discovery 71
Newtonian Separation Meets Quantum Wholeness 73
Scientists Denouncing Astrology 74
Astrologers Debunking Science 74

8. The Resonance Between Astrology and Astronomy 77
Charts for the Discovery of Uranus, Neptune, Pluto, Chiron, Ceres 78-83
Pre-Discovery Observations 84
Determining Orbit Perturbations 84
Elusive and Controversial Neptune 85
Bode's Law and the Discovery of Ceres 88
Science Plagiarizes Astrological Ideas 92
Table of Orbital-Planetary Resonance 94
Collision-Encounter Theories of Planetary Origin 94

9. Searching for Trans-Neptune Reveals Evidence of Transpluto 96
History of the Search for Trans-Neptune 96
Why Projected Positions Are Plus-or-Minus 180 Degrees 99
Pickering and Lowell 100
Accidental Discovery: The Almost Predictions for Pluto 102
Commandments for the Planet-Hunter 103

10. After Pluto's Discovery: The Hunt Continues 104
Astronomers Give Predictions for a Tenth Planet 105
Comparing Predictions with the Current Transpluto Ephemeris 106
"Most Wanted" Poster for a Fugitive Planet 110

11. The Astrological Discovery of Transpluto 112
Table of Hypothetical Planets 113
Astrologers' Predictions for Transpluto 114
Correlating Astrology with Mythology 115
Re-evaluation of Suffering and the "Wounded" Goddess 116

12. Persephone in the Chart and in Our Lives 121
Chart Interpretations: Rape, Divorce, Possessive Mothering 122
Chart of New York Stock Exchange 124
Persephone Portrayed in Children's Books 128
Part-Time Residency 130
Chart of Ashfield, Massachusetts 131

13. The Modern Era: Persephone in Cancer/Leo 136
Table of Pluto-Persephone Aspects 137
The Cancer/Leo Generation Gap 138
Chart of the United States 140
Generations Marked by the Ingress of Outer Planets 142

14. Science, Myth, Astrology, History: The Persephone Connection 143
The 685-Year Cycle of Persephone 143
Table for Bode's Law 144
Pomegranate Seeds, Year of the Gods, and the Cycle of Seasons 146
Table of Ages, Persephone Cycle, Outer Planet Conjunctions 148
Climate Changes over the Centuries 151
Earthquakes, Floods, Volcanoes, and Mass Migrations 152

15. The Powerful Feminine Principle 155
Definitions of the Feminine in Astrology 156
Persephone is Lunar, Venus is Solar 157
Sexualization and Recycling of the Earth 158
Persephone in the Charts of Nuclear, Seismic, and UFO Events 162

16. Discovering a Symbol-Glyph for Persephone 166

17. The Celebration of Ancient Mysteries 176
Eleusinian Mysteries 176
Agricultural and Astronomical Calendars 180
The Wild and Cultivated Grain 182
Pig Worship and Dirty Language 185
Altered Consciousness 193
Vision Quest through Ceremonial Chemistry 195

18. Persephone, the Myth and the Planet 199
Conclusions About a Never-Ending Story 199

Glossary 202
References 205
Persephone Ephemeris 1900 - 2015 235

Foreword

This book is written not with the intention of becoming the authoritative or "final word" on Persephone-Transpluto, but to invite contemplation and open discussion on the topic by presenting a wide range of astrological, astronomical, and mythological perspectives. New discoveries and critical analyses will no doubt invalidate, update, or replace some of what is written here. It is the author's hope that what will remain intact is the book's usefulness as a collection of related information and a reference source on the subject of Persephone-Transpluto, and that the research gathered together here will serve as a springboard for inspired thought. It is with this purpose in mind that information has been extensively footnoted, so that others may continue the research necessary to bring this planet into greater light.

All astrological positions given in this book are geocentric-tropical, unless otherwise noted.

Chapter 1

ASTRONOMY AND ASTROLOGY: REAL SCIENCE VS. REJECTED KNOWLEDGE

People sometimes dismiss astrology by saying it's not "real." This is not completely true. Astrologers use real planets, fixed stars, asteroids; celestial objects whose physical existence can be scientifically validated. These have real positions in the sky which we can point to and designate with either astronomical or astrological terminology. Astrologers also use aspects (angular relationships between planets) and periodic cycles which can be measured mathematically. When astrologers refer to the *Ascendant, the Midheaven, East Point, Vertex,* or *Nodes,* they are speaking of positions produced by astronomical calculation. Although astrologers typically use a different terminology and framework than astronomers, there are numerous ways in which astrology is just as "real" as astronomy.

On the other hand, astrologers also use many "non-physical" criteria, such as *Arabian Parts, Uranian Planets, House Cusps,* and *Hypothetical Planets...* points that have been detected or derived by astrological or psychic methods, and which don't have an astronomical or physically solid "reality."

Astrology is not strictly scientific, but neither is religion, romance, mothering, music, nor many other life endeavors that the debunkers-of-astrology choose not to attack. Scientists, who form a large portion of the debunking movement, judge astrology on the basis of their own criteria for science and consequently dismiss it easily as non-scientific or "pseudo-scientific." Being labeled unscientific in an age of technology means ultimate invalidation, for science has made itself the measuring rod of our era. There is a convoluted logic here that equates "scientific" with *real* or *true.* But we must remember that modern

10

thinking has become quite prejudiced by the scientific bias of our times.

Despite the accusations of paranoid scientists,[1] astrology is not a threat to the progress of science. It simply works differently. Science accumulates and processes data, building its theories upon the observable, and thus perfecting its formulated worldview. Science takes data and creates its theories. Astrology, like other metaphysical disciplines (as well as Mythology), starts with a given framework, a system of symbols which the creative mind can then enhance and enlarge upon. Astrology produces data *from* its theories. It doesn't have to run extensive experiments to prove its theories. The reason that science cannot prove or disprove astrology is that it cannot take astrology's enhanced data and, using the familiar methods of science, work backwards to re-create the original astrological principles.[2] Astrology and Science are opposed only in terms of their directional aim; each one starts where the other ends.

The scientific view and method is mechanical-material-empirical and is based on objective, causal reality. Astrology's view and method is metaphysical-mystical-intuitive and is based on a subjective, synchronistic reality. Science tries to find order in the universe and gathers data to that effect. Astrology already has a map of that order.[3]

According to the dualistic orientation of Western philosophy and our modern science-biased culture, data and theory are defined as either astronomical reality or astrological fantasy. But common sense tells us that the distinction between "exists" and "doesn't exist" is not always so clear, and this realization finds support in the principles of both Quantum Physics and Eastern Mysticism.

We discover that the "border" between reality and fantasy, between "objective" science and "subjective" consciousness, between astronomy and astrology, is actually an artificially constructed one. As we accept a broader, more encompassing, and wholistic perspective of knowledge, such borders dissolve, and we can enjoy greater intellectual freedom.[4]

11

PLANET X — TRANSPLUTO

A good illustration of the artificiality of borders between astronomy and astrology is the so-called hypothetical planet Transpluto. Theories about a planet beyond Pluto have been put forward by astrologers and astronomers alike during the past 150 years. In fact, there is much scientific proof that Transpluto is a real planet orbiting beyond Pluto and that its physical sighting is imminent, merely a matter of time and funding. The story of Transpluto is one of ongoing discovery, and to relate it requires first giving some background from the history of astronomy.

Throughout most human history, all the major planets out to and including Saturn were known because they were visible (to what we moderns call the "naked eye"). Uranus, Neptune, Pluto, and all the thousands of asteroids were discovered only during the past 200 years. Recognizing the physical existence of these objects wasn't possible until we had developed technology such as photography and telescopes. In simple astronomical terms, planetary discovery means aiming instruments at a certain portion of the sky and noticing that one of the dots of light is "wandering" from the regular motion of the background stars (the word planet means "wanderer"). Discovery has often been a matter of accident and "good timing." Uranus was discovered in this fashion in 1781.

There is another way that planets are discovered, and that is by mathematical calculation. The planet Neptune is an illustration of this; it was anticipated and deliberately sought out as a solution to the evident perturbations in the orbit of Uranus. Orbital perturbation is a term from celestial mechanics, the physics of moving bodies, which can be explained by an analogy: Imagine a party with many people milling around, and each one, like a planet, with his own intention to move in a certain path or orbit. (A Cancerian type might be heading for the buffet table, a Taurus is set on stationing on the couch, a Pisces is weaving toward the dancing area.) But when these "moving objects" get near enough to other ones, they get pulled a bit toward them, off their intended paths, drawn into conversation, a shared drink, or some other activity. This pull, in our simplistic analogy of moving celestial bodies, is gravitational force.

Once Uranus was discovered, astronomers could track its orbit and predict a path it should follow if there were no other planets further out in the solar system providing additional gravitational force. It became obvious quite soon after the discovery, however, that Uranus wasn't following the predicted path, so astronomers set to work with their tools of math and physics to figure out where there should be a trans-Uranian planet. This is why Neptune is known as the planet "found on a piece of paper." His official discovery came in 1846.

Readers may be wondering why so much astronomical information appears in this astrology book. The point here is that the planets have physical, astronomical characteristics that are mirrored in their astrological personalities. We can learn more about astrology by interpreting the astronomical data. Coupling these with what we know of mythology, we obtain a truly encompassing view of life.

Take Uranus, for instance. All the other planets rotate around axes that are more or less "up and down" with respect to their orbital path. As ruler of individualistic Aquarius, Uranus has to be the oddball and do it differently; he rotates on his side, with the poles nearly horizontal. Neptune, too, "acts out" his astrological nature, having an orbit that is the most nearly perfect of all the planets, almost a circle. Because his orbit is so "perfect," if another body is passing by, the influence of gravitational force is most noticeable with Neptune ... this is the planet that, according to Astrology, rules idealistic (perfectionist) and impressionable Pisces.

To return to our story, very soon after Neptune was discovered, astronomers noticed that there must be still another planet further out.[5] There were discrepancies in Neptune's predicted orbit and there remained perturbations of Uranus that hadn't been fully accounted for by the pull of Neptune. The hunt began for a trans-Neptunian planet (and the name Trans-Neptune, as we shall see, is really a more accurate descriptor of what is usually called Transpluto).[6]

Ever since the discovery of Neptune, astronomers have experimented with various formulas to determine where a trans-Neptunian planet might be located. The best known formula is based on the Titus-Bode Law (See discussion in Chapter 8). Although their predictions varied

13

as to exact location, many astronomers agreed that a new planet could be found in one general vicinity of sky, namely the area that astrologers call tropical Cancer. When Pluto was discovered in 1930, the first story released to the press was that it had been found because of these predictive formulas. It was gradually realized (though not revealed so directly to the media and the public) that Pluto's discovery was something of a fluke. Calculations determined that in 1930 there should have been "something out there" in the general direction of Cancer, and, sure enough, there was Pluto at 17 Can 46.[7] But when the hoopla over discovery finally died down and the astronomers started re-calculating all the orbits, mass, and real positions of the outer planets, they found that Pluto failed to completely account for the perturbed orbits of Uranus and Neptune.

In other words, Pluto was not the final solution to the formulas, and scientists realized that there must be yet another body beyond Neptune, probably located nearby (in 1930) in Cancer. So the hunt was continued for a tenth planet, and there are still today entire observatories, satellites, and spacecraft involved in the search for Transpluto or *Planet X*. There are currently about a dozen independent programs, including ones at Jet Propulsion Laboratories (JPL) and Lowell Observatory, as well as the examination of information sent from Pioneer and Voyager spacecraft[8] (and, in addition, the investigative work of some astrologers).

By most astronomers' accounting, the evidence shows that a body of varying size exists at a distance of 50 to 100 a.u. (*astronomical units,* or multiples of the distance from the Sun to the Earth). Astronomers calculate that this planet has a very highly inclined and elliptical orbit, and a period of revolution around the Sun of 300 to 1000 years. Compare these figures with those of Pluto, who also travels in an elliptical, highly inclined orbit, but at a distance of 39 a.u., and with a period of about 250 years. Astronomers have theorized that this unknown planet may be related to the formation of comets and asteroids, the chaotic behavior of Pluto, and perhaps be a clue to the catastrophe that brought about extinction of the dinosaur species. They also believe that Transpluto's existence may be linked to long-term cycles of the earth's "wobble" (what we call the Precession of the Equinoxes) and consequent climatic changes on Earth. We must keep these factors in mind; they will come up again later as we discuss the astrological interpretation of astronomical data on Transpluto and the period of its cycle.

Although the physical planet itself has yet to be sighted, accurate calculations have been made that determine probable combinations for Transpluto's orbit, period, and position, which have been converted into an astrological ephemeris. The first ephemeris published for Transpluto's probable positions was produced in 1972 through the combined efforts of the scientist Theodor Landscheidt and the cosmobiologist Reinhold Ebertin (hereafter referred to as *Landscheidt's ephemeris*).[9] Another collaboration of astronomy and astrology produced the better-known and more widely-published ephemeris for Transpluto which is contained in the book by John Hawkins, *Transpluto: Or Should We Call Him Bacchus?* (hereafter referred to as the *Hawkins' Ephemeris*).[10]

According to the calculations used to formulate the Hawkins' ephemeris, the position of Transpluto in 1930 was only 8.5 degrees away from the position of Pluto's discovery point. This proximity plus Pluto's greater visibility explains why astronomers first thought that, with their finding of Pluto, they had discovered the solution to the perturbed orbits of Neptune and Uranus. In the Pluto discovery chart, Transpluto was located at 26 Can 28.[11] Other astrological ephemerides have been produced for Transpluto, and the positions vary only slightly from the *Hawkins* and *Landscheidt* versions.

We have, then, what appears to be a fairly accurate astrological ephemeris showing where Transpluto is likely to be, and astronomers are continuing to hunt for it (its high inclination means there is a greater amount of sky to be searched). We may conclude that Transpluto is not a purely hypothetical planet ... there's too much scientific evidence and funding involved.[12]

Regarding the treatment of a newly discovered planet, many precedents have been set by both the astronomical and astrological communities. There is much that can be anticipated about Transpluto by looking at what occurred during the process of discovering other planets (Uranus, Neptune and Pluto, as well as the asteroids). We shall, therefore, review many facts from the history of planetary discovery, creating a firm foundation upon which we can build our theories.

A NAME FOR TRANSPLUTO

The naming of a planet is always an interesting business, particularly considering how important is the mythology for the astrological associations of a planetary god's identity. All the asteroids and trans-Saturnian planets have entertaining anecdotal histories about how their names were chosen. Various alphabetic designations have been used by astronomers in their search for Transpluto, such as *Planet X* (referring to its status as either "unknown" or as the 10th planet) and Planet O or P (designations used by the astronomer William Pickering). In the 1950s one astronomer, H.H. Kritzinger, was working on the possible orbit predictions for Transpluto and computed an ephemeris; he called the planet *Persephone*. Not only was this the first time a modern astronomer gave Transpluto something other than a generic name, but also this is the name that at least two other astronomers currently searching for the planet say they would use. (Remember, whoever first sights a planet has the honor of naming it.) Both Conley Powell and Mark Littman acknowledge that "Persephone" would be a mythologically appropriate name to give the new planet.[13] They figure that, since Pluto rules the Underworld, the planet beyond Pluto should be his wife-consort Persephone. And at least as of now, there have been no other serious contenders for the name of Transpluto offered by the astronomical community.

Giving names of Underworld characters to objects at the edge of the Solar system is not without precedent, and many astronomers recognize the connection of the outermost planets with the mythological Underworld. There is, of course, the example of Pluto, and in 1978 Pluto's moon was discovered and named Charon, a character from Greek mythology who ferried dead souls across the river Styx to Hades. According to this myth, Charon demanded a ferry toll, so Greek burial tradition stipulated that a coin would be placed in the mouth of the dead.

The naming of a planet is not necessarily a simple, cut-and-dry process, as shown in the example of the naming of Uranus. William

Herschel, the astronomer who discovered Uranus, used his privilege to name the new planet by honoring his benefactor King George III, suggesting "Georgium Sidus." In the 1780s many contemporary astronomers objected to this non-mythological name, as well as to the fact that "sidus" referred to a star rather than a planet. Herschel modified the name to the "Georgian planet," or simply "the Georgian," but correspondence concerning the dispute over this name continued to accumulate and was directed to Johann Elert Bode (an accomplished astronomer best known for his Bode's Law of planetary distances).

Among the mythological names suggested were Neptune, Hypercronus or Trans-Saturnis, Cybele, and Minerva. Bode himself was the one to propose Uranus. This name gained in popularity and the director of the Vienna Observatory (a Jesuit named Maximilian Hell!) adopted Uranus for use in the Ephemerides. Meanwhile, Joseph Lalande, another prominent astronomer, felt that nothing but "Herschel" would be the appropriate name and that "these mythological names are inept today." The three names, Uranus, Georgian, and Herschel, were used interchangeably for sixty years following the planet's discovery,[14]a demonstration of the fact that the *integration of a new planet takes time.*

The discovery of Uranus was not merely a scientific "first." It opened up entirely new possibilities for cosmological and philosophical questions. One of the major breakthroughs brought about by the discovery of Uranus was the idea that, since the boundary of the known solar system was extended by the addition of this new planet, there could be even more undiscovered worlds beyond Uranus. Among those who first pondered such implications was an astrologer, the famed Raphael (R.C. Smith). Of Uranus and the scientific prejudice against astrology he wrote (about 1823), "When we reflect that there may be other planets equally powerful, beyond his orbit, as yet undiscovered, we cannot help remarking the extreme ignorance and folly of those persons, who require from the Astrologer what they expect from no one else, infallibility."[15]

This same Raphael (as well as other contemporary 19th century astrologers) described the astrological characteristics of the "new" planet Uranus, writing that he was almost totally evil in influence, alongside the more objective descriptions of "strange, unexpected, eccentric, original, unsettled, sudden, out of custom, inventive, seeker of novelty and change, associated with astrology, and ruler of Aquarius." It has been a common practice among astrologers to assign malefic influence to new planets (certainly this has been the case with Pluto), but it is just as common for scientific discoveries to be met with initial distrust. Both situations are a reflection of the very human reaction to the unknown: if it's new or we don't understand it, it might or must be "bad."

Science has certainly looked upon astrology with just such an attitude, judging astrology from the standpoint of science. As we hope to show in this book, there is much that can be gained from turning the tables. *We can examine science from the astrological and mythological viewpoint.* From the viewpoint of science, the purpose of astrology is entertainment. We shall see in the following story that, from the viewpoint of astrology and mythology, science can be just as entertaining.

THE NAMING OF PLUTO

Those who are familiar with astrology are quite used to the so-called "coincidences" of life. We find that astronomy is not exempt from such eerie synchronicities, particularly in the activity of planet-naming. As one astronomer admitted, "the history of planetary discovery is rich in drama, in ironies large and small, and in coincidences that even the most conservative of astronomers have referred to as remarkable or incredible."[16] One entertaining example is the naming of Deimos and Phobos, the two moons of Mars, shortly after their discovery in 1877. The names of these attendants to the god of anger were suggested by a man whose own name was Henry Madan! There are too many examples of such "coincidence" to accept them as merely statistical accident.

Perhaps some of the most fascinating stories of coincidence can be found in the naming of Pluto. Although Pluto was discovered in February 1930, the news was not officially announced to the press until four weeks later. The director of Lowell Observatory had decided to make the announcement on the anniversary of the birth of Percival

Lowell (March 13) who had done so much work searching for the planet but had died too soon to witness the discovery. As soon as the announcement was made, suggestions for names began pouring in. Minerva was high on the list of those considered.[17] Other possible names were Atlas, Artemis, Perseus, Vulcan, Prometheus, Hercules, Hera, Zeus, Odin, Freya, Osiris, and Bacchus. One complication was that many of the suggested names (such as Athena and Icarus) had already been given to asteroids. Lowell received one telegraph asking, "Why have only one lady in our planetary system?"[18]

Meanwhile, the following events were happening at Oxford. The brother of the man who had named the satellites of Mars (Madan) was reading the newspaper at breakfast on March 14, 1930, when he saw the announcement of the new planet, and remarked, "What shall we name it?" His 11-year-old niece was at the breakfast table, and pronounced (at 8:05 a.m. GMT), "It might be called Pluto." Prof. Madan thought this such a good idea that he sent a message to a professor of astronomy at Oxford, and from there it passed on to the Lowell Observatory in Flagstaff. A chart drawn for the moment of this naming shows Taurus rising, with chart ruler Venus in fiery Aries. The girl's name? Venetia Burney — burning Venus!

Another interesting coincidence occurred when the president of the Royal Astronomical Society, A.C.D. Crommelin (a world authority on names and discoveries of planets), received the letter suggesting the name Pluto, he also received by the same post a letter from his sister suggesting Persephone. (Crommelin's sister, incidentally, lived in the town of Boars Hill. "Coincidentally," the animal sacred to Persephone is the pig or boar.) He decided, however, that a masculine name was more appropriate.[19] Although an Associated Press article dated March 25, 1930 from Milan told that Italian astronomers were already using the name Pluto, the official naming came later. Director of the Lowell Observatory at Flagstaff, V. M. Slipher, announced on May 1, 1930 that Pluto had been chosen. Note that, in the Pluto discovery chart (page 82), Vesta is exactly conjunct the Ascendant, telling us something about how Pluto was revealed to the world. Slipher's first name? Vesto (a masculine Vesta?)![20] The asteroid was retrograde at the time of Pluto's discovery, and Slipher had unknowingly waited until his namesake planet turned direct before he made the announcement.

Regarding Pluto's rulership of Scorpio (sign of elimination), the following "coincidence" is amusing. One of the Lowell Observatory trustees expressed his reservations to Slipher about the use of the name Pluto. Evidently it bothered him because it suggested a commercial product called Pluto Water, a laxative widely advertised in the U.S. during the 1920s and 30s.[21]

Chapter 2

DISCOVERY OF A NEW PLANET

It is, of course, likely that many of the ideas postulated in this book about Persephone will be confirmed or disproved following the eventual sighting of the planet. The tentative ephemeris will no doubt be corrected, and concepts will need to be refined in order to be consistent with new scientific facts. But much can be learned about a planet's astrological meaning prior to its physical discovery. And there is good reason to believe that the mode of Persephone's discovery will be different from that of the other outer planets. This will become apparent when we delve into the question, "What is the astrological significance of a new planet?"

It is a widely accepted astrological concept that whatever a new planet represents will show up in world-wide events and changing attitudes concurrent with the planet's discovery. In other words, whatever that planet "means" astrologically begins around the time of its discovery to manifest dramatically in the world as a paradigm shift. This is precisely what happened with the earlier discoveries of the three known outer planets. Uranus, planet of revolution and technology, was officially sighted in 1781, concurrent with the American, French, and Industrial Revolutions. Neptune, the planet of illusion and spiritual focus, was found in 1846. The change in consciousness at that time was reflected by the emergence of numerous spiritual and Utopian communities, as well as the invention of photography, which greatly influenced people's concepts of illusion and reality. Pluto was discovered at a time when the world was coming into a new understanding of power, both atomic/nuclear power and the power of dictators.[22]

Humanity in general did not know about technology prior to the discovery of Uranus; his effect was disruptive, especially to people's concepts of traditional social and economic class. They didn't know about photography until the discovery of Neptune; we are quite used to it now, but when it was first invented, the effect on people was confusing (a Neptunian trait).[23] Before Pluto's discovery, atomic power was not generally understood; Pluto's effect was threatening to people's concept of life and death.

There are very large shifts in consciousness when we come into awareness of a new planet, and the effect of this is not limited to changes in scientific knowledge. When we read an old astrology book written before 1930, we notice that the astrological traits now assigned to Pluto were at that time still unformulated or there were bits and pieces of his qualities scattered among the rulerships of the known planets. When consciousness associated with a new planet begins coming into general awareness, the message isn't initially clear, and it takes transition time to become integrated. After all, it has been 200 years since the discovery of Uranus, and humanity in general is still adjusting to the changes attending this event, namely social and technological revolution. Today, sixty-five years after Pluto's discovery, we are still grappling with the issues that he brought in.

CONSCIOUSNESS OF A NEW PLANET:
THE ONGOING DISCOVERY

If we look around us now at all the current changes occurring on the earth, we notice that life on our planet seems to be going through a transformation of major proportions. Many of the changes have to do with a more wholistic awareness of our world; there are threats of greenhouse warming, ozone holes, endangered species and ecosystems, earthquakes, air and water pollution. We also notice that imbalances in the environment are affecting more massive numbers of humanity: hurricanes, climate changes that alter agricultural patterns, epidemics such as AIDS and world hunger, migrations of entire racial groups as well the wandering tribes of the "homeless." Issues are brought to our attention about the current legal and religious status of the right to die (and the right to life or abortion), as well as the freedom of sexual preference. There is a breakdown of Western-style social structures such as the nuclear family and a consequent increase in the numbers of divorces, separations, single parents, adopted and "latchkey" children, as well as the resulting turmoil over "family values." We observe this planetary stress, we ask what it means, we search for some explanation that ties together these apparently separate issues.

Astrological analysis of this situation should inform us that such a dramatic level of crisis-in-consciousness cannot be due merely to the transits of known planets, but must be the reflection of the traumatic birthing of a new consciousness, a new planet ... *Persephone*. As this book shall make clear, the planet and mythological character of Persephone gathers together all of the above-named, apparently disparate problems under one heading, offering a perspective and hope for potential solutions.

Some astrologers have expressed doubts about the reality of a planet beyond Pluto, and have criticized attempts to perceive this planet's meaning prior to its physical existence being determined by scientific observation. But, as we have seen, astronomers already acknowledge that there is "something out there" perturbing the orbits of the known outer planets, and astrologers are sensing that there is "something in here" perturbing our consciousness. Reversing the ancient axiom *As Below, So Above*, the massive inner changes in human consciousness that are taking place today in this world "below" may be a reflection of the emergence of a new planet "above." We need not wait for this new planet's scientific reality to be validated before we apply her astrological principles. Considering the intensity of the world's problems, we don't really have the luxury of that much time to wait.

Just as the discovery of Pluto was a reflection of new energies and challenges for humanity to struggle with, transform, and integrate, so it is with the imminent and ongoing discovery of Persephone. Likewise, there is also the confusion, distortion, and "fumbling" that attends initial awakening to a new planet. We attempt at first to identify the new energy on the basis of old patterns, using methods and language we're familiar with, but we eventually realize this doesn't work, that we must make a leap in consciousness and develop new terminology.

This initial confusion and mis-identification has occurred in recent attempts to understand the new planet astrologically. With so much recent focus on environmental and ecological issues, the tendency has been to identify changes in the domain of "Mother Earth" as associated with the asteroid Ceres.[24] But the Greek equivalent of the Roman Goddess Ceres was Demeter, and who was the Daughter of this Earth Mother but *Persephone?* In the first book published on Transpluto, Hawkins called the planet Bacchus, a Roman God whose Greek name was Dionysos.[25] But who was the Mother of Dionysos? According to many mythological sources, *Persephone!*[26] The confusion around the identity of Transpluto-Persephone has been compounded by the fact that two minor asteroids (the 399th and 26th to be discovered) were officially named Persephone and Proserpina (the Roman name for Persephone).

Just as astronomers were once fishing around for a new planet in the "right" direction and discovered Pluto, but missed the real solution to their calculations, so have many astrologers misidentified Persephone. As we shall see further on, there are numerous characteristics and themes associated traditionally with astrological Pluto or Ceres that actually "belong" to Persephone. We are not here discounting the value of research performed by other astrologers, nor are we denying that astrological Ceres rules earth matters and Pluto the Underworld. We are instead recognizing that some of the astrological principles assigned to Pluto and Ceres need broadening, enhancement, and greater clarification via a new perspective. Once we make the leap in awareness, we can stop limiting our terminology to what we're familiar with, and drop the habit of trying to fit the concept of a new energy into an old philosophical framework.

SIGN RULERSHIP

Once upon a time, all twelve astrological signs were ruled by seven objects: the Sun, the Moon, and the five planets from Mercury to Saturn. The subsequent discovery of each new planet has necessitated alterations in the established system of astrological sign rulerships. Looking at how this came about will give us suggestions on how to "correct" the misinterpretations of Persephone's meaning and identity.

Prior to the discovery of the asteroids and the outer planets, there were double rulerships among the Zodiacal signs. (See Table below) Until the discovery of Uranus, Saturn was the ruler of both Capricorn and Aquarius. Before Neptune was discovered, Jupiter ruled Pisces as well as Sagittarius. Until the discovery of Pluto, Mars had dual rulership over Aries and Scorpio. Prior to the discovering of the asteroids, Mercury had a double rulership over Gemini and Virgo. In each case, the traditional ruler has had to reduce its domain to a single sign and "abdicate" its rulership of the other sign to the new planet. This alteration in rulerships affected the meaning of the signs and planets involved. Prior to the discovery of Uranus, Aquarius was conceived as a much more Saturnian sign, and before Neptune's discovery, the nature of Pisces was more Jupiterian. The quality of both Aquarius and Pisces was expanded by the assignment of a "new" ruling planet.

Rulerships

Planet	Traditional	Current
Sun	Leo	Leo
Moon	Cancer	Cancer
Mercury	Gemini/Virgo	Gemini
Venus	Taurus/Libra	Taurus
Mars	Aries/Scorpio	Aries
Jupiter	Sagittarius/Pisces	Sagittarius
Saturn	Capricorn/Aquarius	Capricorn
Uranus		Aquarius
Neptune		Pisces
Pluto		Scorpio
Asteroids		Virgo
Persephone		Libra

In accepting the new rulerships, we do not abandon the old system entirely. We can still refer to Saturn as the traditional ruler of Aquarius and Jupiter as the old ruler of Pisces. When we speak of Medieval or Babylonian Astrology, we find that the old system is appropriate. But when speaking of modern life, we use the current rulership system that has been expanded by new planets. We are still in a transitional stage with the "new" ruler of Scorpio (Pluto) and the new Virgo ruler (Ceres

25

and the other asteroids) because integration of new consciousness is still taking place. Thus we find recently published astrology texts and articles calling Pluto a co-ruler with Mars of both Aries and Scorpio, and we hear continued debate over the rulership of Virgo by the asteroids. Residual confusion about sign rulership is reflected in the world around us by our lack of clarity about Plutonian issues like nuclear power and personal "empowerment," and asteroid-fragmented definitions of the Feminine.

We need to address this confusion over the rulership of Virgo and its relationship to definitions of what is "feminine." Virgo is characterized by fragmentation. Many astrologers have fallen prey to the rational, "masculine," Mercury-ruled Virgo tendency to over-itemize and "not see the forest for the trees." As a result, a disproportionate amount of astrological influence has been given to individual pieces of flying rock, while the overall Feminine aspect of Virgo has been relatively ignored. The issue of Chiron is a case in point. Many astrologers (this author included) and most astronomers do not consider Chiron to be equivalent in size or influence to the "regular" planets; physically and astrologically he is more like a stray asteroid or satellite, possibly a captured comet. According to astronomical evidence, Chiron is probably a temporary visitor and likely to leave the Solar system at some point. Mythologically, Chiron has long been associated with the constellation Centaurus. The location of this constellation (near Virgo, but off the ecliptic) is no doubt part of the reason that some astrologers have attempted to assign the rulership of Virgo to Chiron.

Anytime we are dealing with Virgo, we must remember to balance the desire to individualize specifics with the need to perceive how those items fit into a larger picture. The asteroids are one large group of individual objects. If Chiron can be said to rule Virgo at all, it is only in terms of his being one of many similar objects. Chiron cannot be said to be the sole ruler of Virgo, anymore than can one individual asteroid, with the possible exception of Ceres, due to her status as being the largest as well as the first discovered asteroid (see Chapter 8). There are many aspects to the nature of Virgo, and healing (which could perhaps be represented by Chiron) is only one of them.

Even if this argument does not persuade, we must give weight to the feminist viewpoint that we really don't need one more masculine ruler in the Zodiac (an idea which stands quite obviously behind this book's gender-identity of Transpluto). Of the feminine rulers of Zodiacal signs, Venus is the only wholly female planet, for the Moon, though symbolic of the Mother, has been de-gendered and relegated to rulership over feelings and subconscious (for men or women), and the asteroids (headed by Ceres) are the divided and scattered elements of a fragmentation of the feminine principle.

Returning to our discussion of general rulership ... When we become aware of a new planet, we not only gain an expanded understanding of the sign it rules, we also develop a new and more mature attitude toward the "old ruler" it has replaced. In the case of Pluto and Scorpio, our growing understanding of power in its nuclear, national, and personal expressions has made us look at war and aggression (the old ruler Mars) in an entirely new light (or, perhaps, darkness!).

After assigning all the known planets to sign rulerships, there is only one set of double-rulership left in the old system, Venus ruling Taurus and Libra. If our logic is correct, the new planet Transpluto must be assigned rulership to one of these signs, which Venus must "abdicate." As we shall see, Persephone will be replacing Venus as the ruler of Libra, and this will mean major transformations in how we understand the meaning of the sign Libra. As a result, we will also experience new attitudes toward the old ruler Venus.

This rulership discussion is based, as is most of this book, on symbolic-mythic reasoning, not a statistical analysis of practical delineation. Readers must interpret the conclusions for themselves, applying what is suggested here in regard to their specific charts. One way, for example, is to look at the house with Virgo on the cusp, and see if Ceres rather than (or in addition to) Mercury seems to rule the issues contained in that house. Likewise, the areas of life indicated by the house with Libra on the cusp can be observed in terms of the new ruler Persephone rather than the old ruler Venus. Using the new rulers will reveal an expanded dimension and perspective on the areas of life and issues related in the houses these planets rule.

27

12 PLANETS — 12 SIGNS

With the additional planet Persephone, we now have twelve planets (including Sun, Moon, and Ceres/asteroids) to rule each of twelve signs, which gives us more than "just" a neat and tidy system of correspondences. It suggests that we have before us the possibility of a truly wholistic view of the Zodiac, somewhat akin to our experience of coming to the twelfth house or the last sign Pisces, and finding a little bit of everything there, all parts aggregating together to form one realization of the whole. The addition of any new planet has tended to alter our understanding of all the signs and planets, but this final addition that completes a 12-to-12 correspondence will undoubtedly give more emphasis than usual to re-experiencing and re-visualizing the entire Zodiac and all the planets in a profoundly wholistic manner.

With a twelfth planet we have the experience of "coming full circle." As we explore the realm of myth, we shall see that this concept of cyclic return "belongs" to no other mythological figure so much as it is does to Persephone.

Chapter 3

REBIRTH OF THE FEMININE:
THE RETURN OF THE GODDESS

The discovery of Persephone is an on-going experience. Astronomers are looking and getting closer; astrologers are seeking to identify her; we are all observing what is going on around us and that will assist us in finding her, both "above" and "below." As stated previously, our interpretations need not wait upon the official announcement of the physical sighting of a new planet. It is entirely possible that the discovery of Persephone will manifest in a different manner than that of the other "discovered" planets. Uranus, Neptune, Pluto are all masculine figures; what their discoveries brought into human consciousness was externalized and yang. The developments concurrent with their discoveries (technology, photography, and atomic fusion) were all products of the rational, masculine mind externalized upon humanity. Their discoveries as planets were hailed as triumphal events of Science (that most masculine sport of mind-over-matter).

The current movement of consciousness into our awareness is not of this masculine sort. It is not descending upon humanity from above, from the mind. The problems of ozone holes, water and air pollution, endangered species, etc., are not the yang action of a male god. What is happening now is oozing up from below, from underneath, from the body of the earth itself — herself, that is. These eruptions are a feminine re-action to too much yang, excessive technology, lack of sensitivity to the needs of the earth and its children who have had their fill of mind-over-matter. Humanity is now experiencing increased awareness of intuitive processes, the dream state, conscious dying, and the so-called Rebirth of the Feminine. But civilization has witnessed several thousand years of investment in the masculine, yang orientation, and the result is an extremely polarized reality. Re-balancing these extremes is not apt to be a gentle experience.

Because there is a current generalized trend of awakening to internalization and feminizing, it seems unlikely that the new planet (as a symbol of this process) would be discovered in the masculine, externalizing mode of scientific proof. It would be more consistent with the nature of a feminine planet to be first discovered internally, intuitively, and astrologically, before the astronomers find her (if indeed they ever do).

We have the challenge of discovering Persephone within ourselves... not with telescopes and formulas (through magnification of the mind), not by looking further and further outward, but by looking inward (through reflection of the feelings). As we venture along the path of internal discovery, we experience separate moments of realization and we recognize separate parts of the puzzle. We seek a framework, something to hold together the various images. And where do we find such guidance? *Mythology:* the stories of the gods and goddesses who represent all aspects of the development of human consciousness. Myths relate the drama of the inter-relating parts of ourselves.

PERSEPHONE AND THE WORLD TODAY

We have determined that whatever a new planet symbolizes or represents will be emerging into general consciousness while the discovery of that planet is taking place. If Persephone is the correct identity of this new planet, then we should be able to observe the elements of her myth manifesting all around us. A quick, overall summary of the myth (in its popular version) goes like this:

Persephone is an innocent young maiden (*Kore,* pronounced core-ray, the Greek word for maiden) and she lives in a Garden-of-Eden existence. She is out in the fields one day, unprotected by her mother (Ceres-Demeter), when suddenly Pluto comes exploding up out of the ground on his chariot. (The Earth Goddess Gaia has agreed to open up the earth to allow passageway for Pluto to capture Persephone.) He carries her off, taking her down into the Underworld to be his Queen. Demeter goes looking high and low and can't find her. She eventually discovers that Pluto has abducted her daughter, with the apparent approval of the other gods on Olympus. Demeter is so angry that she stops everything from growing on the earth, and then wanders off in mourning for her lost daughter. The earth appears to be dying and experiences its first winter. Eden has been disconnected.

The gods on Olympus begin haggling over what to do about the situation. They arrange for Persephone to be returned, but she has already eaten the food of the dead, which means she is condemned forever to the Underworld. After further debate, it is worked out that Persephone will return to her mother in the spring and then go back to Pluto's world in the fall. And so, the earth blooms in the summer because of the Earth Mother's joy at her daughter's return, and the earth is dead in the winter because Demeter mourns her daughter's absence.

The world of today is envisioned in this Persephone myth in the following ways:

A main feature of Persephone's story is the *constant cycling of separation and reunion.* She separates from the earth plane and her mother in order to unify with Pluto. When she returns to her mother again, she is also separating from her partner and her "other" home. It is a simultaneous separation-reunion. We see this reflected in today's high divorce rates and legal hassles over child custody, the millions of single mothers, the current transformation in partnering and parenting, the revitalized interest in mother-daughter relationships. The symbolism of the child abducted from the protective, nourishing mother is even shown to us in the current practice of putting photos of kidnapped children on milk cartons! Unprecedented numbers of children are currently being abducted by one of their separated or divorced parents. New and stringent laws are being enacted to enforce the rights and protection of single mothers, as well as the responsibility of "deadbeat dads." There is experimentation with "open adoption," and media bombardment of stories about adopted children, now grown up and seeking out their biological parents and siblings.

The myth also relates how *Persephone lives in two worlds, walks the borders, mediates between them: the worlds of conscious and unconscious,* the worlds of life (ruled by her mother Demeter-Ceres) and of death (ruled by her abductor-husband Pluto). We notice this reflected today in the use of mind-altering drugs as well as the various techniques of meditation, past-life regression, rebirthing, dream therapy, "conscious dying"[27] — all aimed at traversing levels of consciousness and investigating paranormal experience. When Persephone returns from the Underworld, she brings back the knowledge of the treasures that Pluto guards. Appropo of this we note the popular interest in near-death experiences (NDEs), and the widespread public concern over the misuse of consciousness-altering drugs and alcohol. We also see that legal and religious authorities are struggling with the issue of conscious choice of the right to die (or "murder" a fetus), as well as with the interpretation of the "living will." There has also been the development of Hospice Care and court debates over "assisted suicide."

The myth of *immortality* and the notion that one could die and yet "live to tell the story" has actually fascinated mankind for a long time. The mythologies of every known culture feature some version of this ability to temporarily visit the Underworld. What is pertinent to our argument here is that there has been in recent times a conscious focusing on death from a perspective other than the religious viewpoint. In a sense, the astrological (and mundane) "rulership" of death has moved retrograde in the Zodiac, from Saturn-Capricorn (the ancient ruler), to Sagittarius (organized religion), to Scorpio (organized medicine). Most recently, we experience death as ruled by the legal profession (Libra), because ever since Pluto's discovery, the "miracle" of medical technology has posed moral questions about lengthening the life span and the artificial delay of death, areas that the legal authorities have pounced upon with enthusiasm. Needless to say, the process of aging (Saturn) still goes on, so despite our wonderful and expensive technology, what we get is an extension of the aging process, and thus a larger population of older people with everyone being "older" for a greater proportion of their lives. No matter how much power that religious institutions (Sagittarius), medicine (Scorpio), and lawyers (Libra) proclaim to have over our behaviors and attitudes about mortality, Saturn still rules aging and death.

To return to the Persephone myth, the combined participation of Demeter-Ceres (Mother Earth), the earth goddess Gaia-Ge, and Pluto means that the story is also about *the earth in upheaval with ecological disaster.* We see this manifesting all over the planet in the form of earthquakes, volcanic eruptions, and climate disturbances, matched with a simultaneous growth in whole-earth consciousness ("Save the Whales, Rainforests, etc.") and serious efforts at recycling.

Within Persephone's story and her relationship with Pluto are questions about the *imbalances of masculine and feminine forces;* it proposes an androgynous solution. This is reflected in the current concern for Women's Rights, Gay and Lesbian Rights, and related issues of Equality. We observe the downside projection of this myth as AIDS and the high statistics of rape. Persephone's partnership with Pluto tells of the possibility of the re-definition of relationship as *co-independence;* we see this mirrored in the vast numbers of self-help books and workshop trainings focusing on resolving co-dependency and abuse, and developing more conscious ways of partnering. There is much pop psychology centered around the idea that maybe marriage, partnership, and sex can revolve around something more conscious than procreation.[28]

We also notice by association the involvement of *Persephone's son Dionysos* (the wild god of vegetation, wine, dance, and ecstasy) in the current Men's Movement, the "Green Man" and Iron John of Robert Bly fame, and the organized groups of "white guys with drums" seeking emotional expression and "getting in touch with feelings." We also observe contemporary questioning of whether a man's behavior is really determined solely by two factors, either his head or his penis.

In the story of *Persephone's abduction by Pluto, the "visitor" from another world,* we can even see a reflection of the current increase in reported UFO contacts and alien abductions. It is interesting to note, in this regard, the involvement of Persephone's mother Ceres, ruler of agricultural crops, with the appearance of *Crop Circles* associated with UFO landings. One of the leading investigative groups studying this phenomenon is named CERES (Circles Effect RESearch). The story of Persephone emerging from hiding into the light of day is being enacted as the Plutonic secrecy of government cover-ups about UFO sightings is now being revealed.

During the past two decades, many voices have been announcing what has been called the *"Return of the Goddess."* The assumption has been that the Goddess is now returning to earth from the Underworld. But if this were really a matter of Persephone simply returning to earth, would not her mother be rejoicing and making the earth green and wholesome again, as in the myth? Surely, there has been increased concern and attention given to healing the imbalances in the earth's atmosphere, oceans, wildlife kingdom, the entire eco-system, but the disturbances are manifesting on catastrophic levels, and this is not evidence of a mother celebrating her daughter's return. It seems more like the portion of the myth in which Mother Earth-Demeter-Ceres is mourning the separation from her daughter. We might ask if Persephone is, as usual, doing her Libran "some of each" number, and is both returning to earth and simultaneously returning to the Underworld (the unconscious) to perform the work that is needed there. As Persephone descends to re-join Pluto, humanity has the choice to either "go down" (as in the expression "go down with the ship"), or to descend into greater depths of inner understanding, eventually cycling up again with re-emerged awareness.

Persephone crosses borders, dissolving assumptions about the limitations of boundaries. Her travels between worlds is reflected in the recent scientific advancements that have brought about space travel and turned science fiction into scientific reality. We see this in the new images of popular culture such as Star Trek ("to boldly go where no one has gone before"), which features transporter beams and worm holes as means of travel. We also cross into alternative worlds with the new developments in computer technology that are giving us "virtual reality."

Persephone's story includes *the wandering of a mother in search for her daughter.* This, along with the wandering of Persephone herself (from her earthly home to her subterranean one and back again), is highlighted today by the plight of the "homeless" and massive populations of refugees. It is also symbolic of the fact that more people than ever before in history are migrating to new homelands and from rural areas to cities.[29]

It is truly astounding how accurately the Persephone myth relates what is going on in our world today. The story is traumatic and stressful, yet it is not without a hopeful note. Persephone's myth tells of the eternal return, the carrying of the seed of renewal, the promise that life goes on, even after death and separation. It is truly a myth for our time: the Heroine's Journey.

Chapter 4

MYTHOLOGY AS MYSTERY
OR
THE MIS-STORY OF HERSTORY

Mythology is the key to enlivening bare principles of astrology with meaningful analogies. For our purposes it serves as a non-rational "proof" that we are correct in identifying the new planet Transpluto as Persephone. Greco-Roman mythology has always been an accurate source for identifying the previous "new" planets. It is consistent with tradition to look to mythology again for inspiration.

But myths are expressions of their chronological age within the history of civilization. What has "come down" to us has been translated and filtered through the sieve of our own cultural prejudices. The result is that the stories that are passed down are often diluted and sanitized versions of the original tales. We must remember, in dealing with myths about goddesses, that our culture has witnessed 4000 years of development of the Father-Son-Holy Spirit complex. Prior to that development, culture was centered around the Mother-Daughter-Old Wise Crone matrix. But what about even earlier periods?

There are some indications from archaeological study that the Persephone-Pluto myth, or something very similar to it, may be one of the oldest stories in the world. The ancient cult of the Mother Goddess is commonly regarded nowadays as a concomitant of humanity's transition from hunting and food-gathering to agriculture. It is true that the dependence upon the fertility of the soil, the importance of sowing, growth, and harvesting are all reflected in the concept of the Earth as the Great Mother. But ancient peoples did not have to wait for the invention of agriculture to be overtaken by the realization of woman as the spring of life. Although the mythology of the Great Mother fits well with agricultural symbolism, there is little actual evidence (apart from Eleusis, many thousands of years later) for its specific connection with agriculture.[30] What evidence does exist seems to point to matters more profound and cosmic, namely the relationship between life and death.

What archaeologists have found at Catal Huyuk, an excavated site dating back some 8000 years, shows that the belief structure spoke of the communion between the living and the dead. Bones of ancestors were gathered and buried under the houses of the living, as was fairly common among Neanderthal Man. The images at Catal Huyuk show that the horror of dying remains unabated, but the power of life is constantly regenerated and embraces this terror. Our interpretations tend to make abstracts out of the crude symbolism of the ancients, speaking of the "concept of fertility" about amorphous breasts that protrude from walls, or the "symbol of death" about depictions of vultures feeding on flesh. But one thing is clear: there is the presence of the Goddess as the eternal source of Life, the unending miracle of birth, and there is her perennial antagonist, Death. The identification of this prehistoric goddess is closer to what we understand in Greek mythology as Gaia, unconscious creativity. Her transformation into Mother Earth ruling agriculture came closer to historical times.

To ancient man, Mother Earth was mistress of life and death. Death was not a separate enemy; death was part of herself. All life returned to her and was reborn through the seed. Pre-Hellenic cultures understood the Goddess as double. This is why we find that Demeter and Kore were spoken of as one being, which was a unique occurrence in Greek mythology.

Many of the myths which survive today originated at the time of the transition from the old matriarchal structure of religion and society to the patriarchal framework we're so familiar with now. This transition time dates to around 2500—1500 B.C., when the great Precessional Cycle was moving from the feminine Age of Taurus into the masculine Age of Aries. The earliest reference to Persephone, an illustration on a Minoan cup (see below), dates to about 2000 B.C. It is a widely accepted idea that the myths which originated during this transitional time reflect that era's takeover by patriarchal forces, and this is why there are so many stories of male gods conquering and seizing wives ...

Earliest known depiction of Persephone and the Narcissus flower (Minoan Crete, ca. 2000 B.C.).

in this way the ancient goddess tradition was absorbed and the goddesses themselves were relegated to somewhat inferior positions within a revised mythology. Since much of our knowledge of ancient history is based on written records, and earliest writing coincided with the patriarchal shift, our traditional understanding is truly "his"-story.

One example of this patriarchal "retrospective conversion" is the adaptation of Persephone's myth which is now in current vogue. Few people realize that the best-known feature of this modern version, *The Rape of Persephone,* is an event that did not even occur in the original story. The version commonly accepted today is very late, historically speaking, and reflects the patriarchy's adaptation re-written to suit its purposes. It is important for psychotherapists, astrologers, and anyone using mythology for enlightenment purposes to recognize this problem, to learn to distinguish the age of a myth, and to dig back to find the most ancient tellings. Although one could argue that modern myths are a true reflection of the culture-at-hand, and that therefore we should use them just as they are, there is something to be said for "cleaning up" a myth tainted by the modern inclination to romanticize. The popular penchant for the "Disney version" of life tends to obscure the more profound and raw experience that is expressed in ancient myth. When we study the older versions, we not only achieve deeper revelations, but we also avoid pitfalls and traps of patriarchally-induced mis-interpretation. One astrologer recently interpreted the Rape of Persephone as a situation where "the daughter is forcefully abducted from the control of the mother," saying that this explains why feminine individuation "doesn't occur easily or willingly."[31] Such is the likely and limited conclusion if one reads only the latest (Rape) version of this myth.

A little research will show, as Charlene Spretnak confirms in *Lost Goddesses of Early Greece,* that "there was no mention of rape in the ancient cult of Demeter and her daughter ... It is likely that the story of the rape of the goddess is a historical reference to the invasion of the northern Zeus-worshippers, just as is the story of the stormy marriage of Hera, the native queen who will not yield to the conqueror Zeus."[32] Spretnak points out that the portrayal of Persephone as a rape victim was added after the societal shift from matrifocal to patriarchal. In the earlier myth, Persephone goes of her own volition into the Underworld. Recognizing that the female archetype goes willingly and consciously among the dead sheds an entirely different light on the matter of "feminine individuation" than does the victim-consciousness of the later version.

Most of us are familiar with Persephone's myth in its modern, sanitized version as an explanation for Nature's change of seasons. This is not, however, just a sentimental story of an innocent girl raped and then returned to the waiting arms of Mother. Close reading of the early and pre-patriarchal versions will reveal much more depth and complexity. Such an examination not only offers potential for altering and improving the current views on psychological development of the female, it also suggests a new approach to social psychology. The modern tendency is to sympathize with and exalt the victim, and to blame and punish the victimizer. We consequently have developed an entire social services system burdened by competition over who is the most victimized (and therefore the most entitled to services) .... the disabled, the endangered, or any particular class of people identified as at-risk by race, gender, age, or income.

The inherent problem is that victim-consciousness is a perfect justification for the human vice of greed. Every time we define yet another category of neediness (lack or loss which requires "fixing"), we can observe acceleration in the clambering for a piece of the pie. An example is the creation of the Americans with Disabilities Act, which released millions of dollars for potential support of equalization. No sooner was this law enacted but 49 million Americans (including drug addicts) suddenly claimed to be disabled. Such a social measure could be initiated as an attempt to legislate an "ethic cleansing" to eliminate prejudice, but because it creates a class *entitled to more on account of need,* it perpetuates a dynamic of assuaging guilt with entitlement-bestowing. When the culture defines people by how much they don't have (loss-lack-victim), it encourages the insatiable desire to appear lacking and therefore entitled to get more. In a values-obsessed society like America, it then becomes politically correct to invent proper vocabulary and behavioral rules that exalt whomever is (according to the latest fashion) oppressed, threatened, endangered, dis-empowered, at-risk, or wounded.[33]

Something is very wrong in this picture. As we try to make sense out of disruptions and changes in our system of value-priorities, we might look to the myth of Persephone and her supposed role as victim *nonpareil.*

TELLING THE STORY

In order to fully explore the meaning of the Persephone myth and its implications for astrology, we shall learn in this book as the ancients did, by telling the story over and over. The original myth dates to a time prior to the invention of writing, when knowledge was passed on by oral tradition. The ancients learned the lessons of their stories just as young children do, by rote and repetition. We, too, shall review the myth repeatedly throughout this book, each time examining more details and filling out the story with more characters and nuances. We must keep in mind that each detail, no matter how minute, is vital to the meaning. There is a reason for every aspect related in the story.

In most of the popular books on mythology (including children's books), all we get is a quaint folktale to explain the change of seasons.[34]   But there is much more to the story than such a simple Hollywood-style plot of Persephone the helpless victim, Pluto the evil violator, Demeter the angry mother-in-law, and everyone else wringing their hands until the "happily-ever-after." If we are going to discover this planet within ourselves, we must examine the myth more closely and turn to the more ancient versions. One place to start is with two popular psychotherapy books, Jean Shinoda Bolen's *Goddesses in Everywoman*[35] and Jennifer Woolger's *The Goddess Within.*[36]   There are numerous publications like these, still in print or available in some public libraries, which aim to present female archetypes or personality types in such a way that offers women keys to self-understanding. Nearly all such books, excellent as they are, rely for their interpretation of Persephone on one same text, the Homeric *Hymn to Demeter*. This text has come down to us through several translations and modifications which date as late as the 8th-to-6th century B.C., long after the "heyday" of the matriarchy, and therefore (from our standpoint) suspiciously colored with too much patriarchal prejudice.

FEMALE VENUS-ENVY

Even with this drawback, Homer's *Hymn to Demeter* can be used as a springboard to deeper understanding, especially if we allow curiosity to lead us into important questions: Why was Pluto so attracted to Persephone in the first place? We learn from other sources that this is because he was shot with an arrow sent by Eros, the servant of Venus-Aphrodite. Why was Venus so interested in making Pluto fall in love with Persephone? Again, when we investigate other myths, we find that this was because she wanted to "get back" at Demeter-Ceres, who had made derogatory but very accurate remarks about Venus, slighting her power. This was a case of "I'll show her! Demeter thinks I don't have any power, but I'll cause the death-master to steal her dearest possession, her daughter."

Not very nice behavior for the so-called Goddess of Love? As stated previously, awakening to Persephone, particularly as the proper astrological ruler of Libra, is going to change our attitude about the traditional ruler Venus. What comes as an astounding revelation is that this new rulership was even foretold by Demeter within the myth itself. The statement Demeter made that had really upset Venus the most was her prophecy that one day, Venus would be replaced by her daughter Persephone!

Another "negative" remark by Demeter was a reference to Venus being a recent upstart, a latecomer (to the Greek pantheon), yet this description was a commonly used epithet[37] for Venus and therefore "yesterday's news" to the Greeks. In fact, the arrival and incorporation of Venus-Aphrodite as a revered goddess occurred very late in the development of Greek mythology. (Few centers of Aphrodite worship have been excavated, or even found, and the pre-Greek language Linear B shows "no trace of Aphrodite.")[38] It is no surprise that she appeared on the scene in Greece concurrent with the shift to patriarchy and the dis-empowerment of the much older goddesses such as Demeter. We will soon see that the myths of Venus reveal certain negative aspects of her personality as the projection of unevolved masculine assumptions about women, as well as an image of relationship that is really quite limited to the realm of bio-chemistry (sex hormones). For now, it is enough to observe that Venus, like Eve of Genesis, was not a matriarchal-based creation. She was, and still is, in many ways a man's (mis-)conception of a woman.

This male "conception" is quite literally portrayed in the myth of her physical conception. The Birth of Aphrodite has been described poetically as "born of the foam," and she is pictured euphemistically as rising from the sea. Historians have interpreted this as the arrival of the Cult of Aphrodite as a cultural import into mainland Greece from across the sea. The myth itself states that it all began when Saturn overthrew his father Uranus, castrating him and tossing his genitals into the sea, which then foamed up. From this demasculation was Venus born, which explains yet another of Demeter's demeaning remarks, that Venus was "the one without a mother." As a statement coming from the Earth Mother (super-mom-of-them-all), this was indeed the ultimate put-down.

Considering that Demeter and her Cult followers used very coarse language (this is confirmed by research into numerous ancient sources), we can imagine that Demeter's original reference to Venus may have been something like "daughter of a prick" (the feminine counterpart to the male-sexist phrase "son of a bitch?").

Lest we alienate some readers with any further profanities, perhaps an apology or disclaimer is in order on two points. (1) As we proceed with the story of Persephone, there is no deliberate intention to be offensive; we are merely relating the actual facts about ancient culture and religion, "telling it like it is" (or was). It is common, among the most ancient and primitive cultures of the world, to find fertility deities using crude language and making baudy jokes about the "facts of life." Ancient Greece and its cult of Demeter-Persephone was no exception.

(2) It may appear in this book as though we are portraying Venus in an undeservedly bad light, that we are "robbing Venus to pay Persephone." We must reiterate that the distinctions we are making are based on mythological, not applied, astrology. Our purpose here is to build a more realistic picture and to scrap the disproportionate and idealized view of Venus which has been adopted by astrologers and which lacks an ancient mythological basis. Modern Western astrology has called Venus a "benefic." Yet the myths of Venus portray her as jealous, possessive, vain, and downright dangerous. It is time to realize, as the Mesoamerican and Mesopotamian cultures did long ago,[39] that the astro-mythological significance of this planet is far from benefic. Such a reputation is not solely a reflection of patriarchal misogyny.

Demeter's prophecy of the replacement of Venus by Persephone is fulfilled in psychoanalytic theory. Venus was insulted that her power (based on unconscious, biochemical attraction-repulsion, jealousy and possessiveness) was questioned by someone older, wiser, more mature, and more enduring (the Mother-Creator Demeter). Her reactive, "get-back," vengeful nature makes Venus a passive-aggressive instigator, and she orders her go-fer and lackey (Eros) to send an arrow that sets the whole Persephone myth in motion ... a myth that will eventually reunite those very opposites upon whose separation Venus depends for worship and self-exaltation. The old fear-based "battle of the opposite sexes" and glorification of sex differentiation, epitomized by Venus, plants its own seed of dissolution.

Chapter 5

PERSEPHONE'S RULERSHIP OF LIBRA: THE UNION OF DUALITIES

One of the fundamental analogies used in Astrology is that the birthchart captures the moment of birth rather like a photograph, a frozen moment in time and place. Birthcharts do "work," but not in terms of this illusion. Real birthing is a *process,* not a fixed, finite event. A good illustration of this birthing process is the current, on-going discovery of Persephone. In the metaphysical sense we are all parents, midwives, and attendants. As a physical planet, Persephone has not yet "crowned," but once she is physically sighted, we will have her official birth certificate (date, time, place of discovery). In the meantime, each of us can know she is already born into consciousness.

An examination of sign rulerships will give us deeper understanding of this emerging planet. What is suggested here is that Persephone fills in some of the gaps that appear in the traditional theories of sign rulership. In traditional, standard astrology there are ten major planets (Sun, Moon, and Mercury through Pluto). If we count the asteroids as one united, eleventh "body," the arrival of Persephone brings the total number to twelve planets, correlating with the total number of Zodiacal signs.

Many esoteric sources and some ancient astronomical texts refer to there once being only ten signs of the Zodiac. Leo and Virgo were once combined in one sign, the Sphinx, a woman (Virgo) with the body of a lion (Leo).[40] Likewise, the constellation Libra was once the "Claws of the Scorpion." The division that occurred with these signs was emblematic of the polarization which was necessary and functional to the evolution of human consciousness. As the Sphinx divided into masculine (fire) authority of Leo and the feminine (earth) temple of Virgo, so did the feminine (water) authority of old Scorpio split off part of itself to become the masculine (air) objectivity of

Libra. In seeming contradiction to division, the power of each original realm was actually multiplied for the purpose of greater interaction and further growth.

The reality is that the Universe, or life, is essentially whole, but divided. Our confusion comes when we assume that divided means separated. To the Western mind, division by two tends to imply polarization into opposites. And when we think in terms of opposites, we often use the handy image of male-female, typified by Mars-Venus, who represent highly polarized expressions. This anthropocosmic approach is typical throughout the world. It is the idea of life that humans project upon the cosmos that *the world is sexualized.* This view embraces more than sexuality itself; it is a general concept that everything, from plants and animals to stones and tools, is endowed with a gender. Many languages assign a gender to nouns. The Mesopotamians divided metals into male and female, and even today, jewellers distinguish the sex of diamonds according to their brilliance. Arabian mystical writing declared that "romantic love is not peculiar to the human species, but permeates all things, heavenly, elemental, vegetable and mineral."[41] The most transparent sexual symbolism is to be found in the images concerned with the Earth-Mother, so that caves or mines are compared with the womb of the Earth-Mother, thus the ancient belief in the gynecomorphic birth of ores. Sacred rivers were thought to have their source in the generative organ of the Great Goddess (the Sumerian word is the same for "river" and "vagina").[42]

So astrologers are not unique in assigning gender to signs or planets. Mars is unequivocably a male planet, ruling the masculine sign Aries ("masculine" meaning yang, extroverted, odd-numbered, positive polarity, fire-air quality). Venus is definitely a female planet, ruling the feminine sign Taurus ("feminine" meaning yin, introverted, even-numbered, negative polarity, earth-water quality). What we must look at, however, is the modern Western tendency to define extreme polarities as separate and opposing, an approach that denies the experience of life as unified and whole.[43]

Pluto and Persephone are not polarized in the same way as Mars-Venus. Pluto rules the "feminine" water sign Scorpio, while Persephone rules the "masculine" air sign Libra. Their purpose is not to polarize in direct opposition, but to function in the way that "opposites" can balance each other and cooperate for unification, almost like a mutual reception. Due to this polarity reversal, Pluto has long been confusingly labelled a feminine planet, and Persephone has been mistaken for her male child Dionysos (Bacchus).[44] Both Pluto and Persephone challenge the established view of men and women as polarized beings; their partnership elicits a growing awareness of androgyny — the *animus* within woman and the *anima* within man. Together they work to unify self- and other-consciousness, a state of being which is more enlightened than the Mars-Venus exaltation of "the opposite sex."[45]

The astrological rulership of a planet should always make sense in terms of the mythology of the particular god or goddess for whom the planet is named. In the myth of Persephone, she descends each year to join her husband Pluto, and rises again in the spring to rejoin her mother Ceres. It follows that Persephone should rule the Descendant, the horizon, the border between day and night, the razor's edge between the earth (Virgo, ruled by Ceres) and the underworld (Scorpio, ruled by Pluto). In a "Libra-rising" chart for Persephone, she would symbolically "rise" to enter Virgo, and "descend" to enter Scorpio. (See figure below)

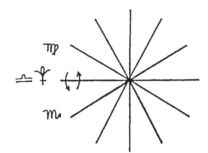

Persephone's Libra-Rising Chart shows movement across the Ascendant-Horizon, representing the connection of the unconscious 12th house and the conscious 1st house.

The similarity of the symbols for Virgo and Scorpio has caused speculation about whether there was once "some connection, now forgotten, between Scorpio and the formerly adjacent Virgo."[46] Persephone seems to fit the bill as a natural linking mechanism, and Libra is the sign of the mediator. Persephone works as a Libran equalizer to even the scales. She was called by the Greeks the *Judge of the Dead* (the "dead" referring to the unconscious), while Saturn is the judge of the conscious (five-senses world of consensus reality). We recall that Saturn is exalted in Libra, an appropriate sign for matters of weighing, judgement, objectivity.

The table below shows a proposed system of planetary rulership that takes Persephone and the asteroids (headed by Ceres) into account, filling the missing links that have existed within the traditional system. There has been much published material discussing the various merits of assigning rulership, but none that include Persephone, and none (we believe) that address the issue of mythological validity as completely as the following system does.[47]

	Rulership	Detriment	Exaltation	Fall
☉	♌	♒	♈	♎
☽	♋	♑	♉	♏
☿	♊	♐	♒	♌
♀	♉	♏	♓	♍
♂	♈	♎	♑	♋
Ceres and the other asteriods* ⚳•	♍	♓	♊	♐
♃	♐	♊	♋	♑
♄	♑	♋	♎	♈
♅	♒	♌	♏	♉
♆	♓	♍	♐	♊
♇	♏	♉	♌	♒
Persephone; author's symbol** ⚵**	♎	♈	♍	♓

We notice how the drama of the Persephone myth is symbolized repeatedly in this system. The myth begins with Venus feeling threatened by the power of the goddess Ceres (Venus "falls" in Virgo, the sign of Ceres' rulership). Venus sends an arrow into Pluto's heart, causing him to fall in love and abduct Persephone. (Pluto is in detriment in the sign that Venus rules.)   Venus appears to triumph over the old Mother goddess by causing her daughter to fall or descend into Hades. (Pisces, where Venus is exalted, is where Ceres is in detriment and Persephone "falls.") But there is rejoicing (exaltation) when Persephone returns to the place of her mother's rulership (Earth-Virgo), and Persephone is, indeed, exalted in her mother's sign.

Further correlations with the myth are found in Gemini and Sagittarius, signs which have been lacking assignment of "fall" or "exaltation" in many traditional systems of rulership. Jupiter plays an important role in the myth as the one who colludes with Pluto to plan the abduction that causes Ceres to go into mourning. (Jupiter rules the sign where Ceres is in fall.) Later on in the myth, Jupiter sends Mercury as *Psychopompos* to Hades to escort Persephone back for a joyful reunion with Ceres. (Mercury rules the sign where Ceres is exalted.)

The presentation of this system of rulership will hopefully clear up some of the confusion which has arisen around the intermingling of the characters involved in Persephone's myth. Hawkins identified Transpluto as Bacchus (Dionysos), and identified the planet as ruler of Taurus. It has been something of a tradition to associate this earth sign with feminine planets (Venus is of course usually given rulership of Taurus, the Moon is exalted in Taurus, the asteroid Ceres was discovered in Taurus), and it could be said that the process of growth is represented by Taurus. But Pluto is in detriment in that sign, and as such he expresses random growth (as in random mutation from radiation exposure), or destructive growth (as in cancer and immune-deficiency). Pluto without his Queen can manifest extreme conditions. Persephone rules the sign of moderation. She *directs growth* through the maintenance of balance between dualities.

Venus (in detriment in the sign Pluto rules) and Mars (in detriment in the sign Persephone rules) are the disguise or masquerade of "separation polarity," while Persephone and Pluto are dance partners on another level of relationship. As supposed ruler of Libra, Venus has been acclaimed as holding the power for what Persephone truly represents. Venus has been portrayed as holding a mirror, assumably admiring herself.[48] But it is Persephone, the true ruler of Libra, that is the mirror symbol for the Soul's reflection. Pluto takes us into the experience of death and we become lost like the souls in Hades, crying out for redemption. But by recognizing deeper meanings for self-reflection and relationship ("other"-reflection), we bring Persephone forward, and she releases the containment energy of Pluto, offering us a measure (via the Scales) of our consciousness, and insight (via mirroring). Through the ongoing deaths and rebirths of our lives, our awareness that we are whole is expanding. And like the lost souls of Hades, we awaken to the light of the Judgement Day and welcome the Goddess.

THE LIBRAN PLANET

Returning to our examination of the myths of Persephone, we observe that she rarely appears alone, but is usually accompanied by the "significant others" of her life ... her mother Demeter, her husband Pluto, even her "rival" Venus. This is another confirmation that she rules Libra, the sign of relationship as well as the seventh house of open enemies. Because she is so closely intertwined with these other characters, and does not really act independently of them (at least not in the myths), we can see how confusion about her astrological identity has arisen. How like a typical Libran personality (and like a woman, we might add), to be known not as herself, but by whom she relates with (she is Demeter's daughter and Pluto's wife). We can understand how astrological Persephone has become enmeshed with astrological Ceres, Pluto, Venus, and the signs they rule: Virgo, Scorpio, and Taurus.

Astrology has said that Pluto rules death-and-rebirth. This broad concept needs clarification and refinement. Pluto does provide the fertile medium (the compost), but it is Persephone as the female aspect that carries the seed. Life cannot be renewed and rebirth cannot occur unless both are present. Ever since Pluto was discovered and the Atomic Age began, humanity has experienced great despair over the nuclear threat. Here we have in our hands a tremendous tool of power, capable of great destruction or great creativity, but we cannot properly use it until we understand what to do with the waste product (the compost!). This is a truly Scorpionic dilemma. Pluto without his partner is a rather morbid, lonely, and estranged character; he is Death without the complete means to achieve Rebirth on his own. Scorpio is extreme; it needs the balancing effect of Libra.

There are all sorts of reversals and counterbalances present in this Death-Dance Partnership of Pluto and Persephone. Scorpio is a water sign and therefore elementally feminine, yet it is ruled by a male god. Libra is a masculine (air) sign, yet ruled by a female goddess. There is much potential to be found here concerning androgyny and the animus-anima interchange. Opposing signs to Scorpio and Libra are Venus-ruled Taurus (feminine earth) and Mars-ruled Aries (masculine fire). One of the best arguments for Persephone ruling Libra is her sequential order in the ancient astrological figure called *The Ladder of the Planets.*[49] (See figure below)

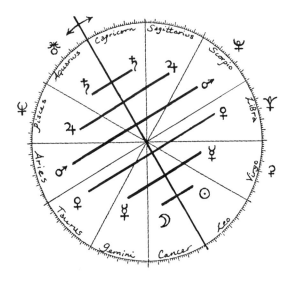

The Ladder of the Planets: the traditional, "old" rulers of the Zodiacal signs, prior to the discovery of Uranus, Neptune, Pluto, etc. The rungs of the ladder connect the traditional co-rulerships, moving upward clockwise from the Moon and counterclockwise from the Sun. As new planets have been discovered and assigned rulership, they have moved downward from the top of the rung (Saturn), and are shown outside of the chart.

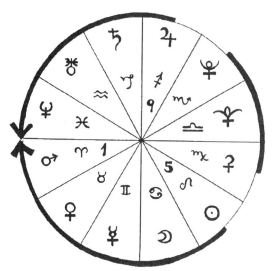

Modern rulership: Personal planets are placed clockwise from the Sun through Mars; the outer planets Jupiter through Neptune move counterclockwise. The emphasis for beginning and ending these two arcs is on the fire triangle (1st, 5th, 9th signs). The asteroids, Pluto, and Persephone occupy the signs of preparation, interaction, and consummation of Relationship.

Libra's rulership by Persephone is a strong one, considering that much of our awareness of Self arises from the mirroring of the Descendant experience. Relationship is where what was unconscious within ourselves manifests and externalizes. In the house of the setting Sun (the Sun is in its "fall" in Libra), the will is sublimated to the desire for union.[50]

THE HIGHER OCTAVE OF VENUS

With the two pairs of opposite signs, Taurus-Scorpio and Aries-Libra, we have a good illustration of the astrological concept of higher and lower octave planets. The basic idea is that planets nearer the center of our solar system (the Sun, Mercury, Venus, Moon, Mars) represent first stages in human development and are called the "personal" planets; while planets further out represent social involvement (Jupiter and Saturn), and the outermost, trans-Saturnian planets are called "transpersonal." The further out we go in the solar system, the more the planets express transformation, complexity, and potentially more evolved levels of what is first experienced by the inner, personal planets. The inner and outer planets have been paired up by astrology, so that what is operating on a private level with a personal planet is expanded to a transcendental and inspired level by its "higher octave" planet. There are several such octave-pairing systems in Astrology.[51] The following explains one system that seems appropriate to the discussion of Persephone.

Mercury is the mind, and Uranus is the higher octave of this mental plane, representing the *idea* of transformation. Uranus is present when we first get the idea of change; we have a mental or rational understanding of a transformation. The Moon is feelings. Neptune is a higher octave of this emotional plane; he is present when we understand change and the *feeling* of transformation. Mars is action and physical experience, and Pluto is the higher octave of this physical plane. He is the actual *experience* of transformation. When people say they're "really going through a heavy transformation," that's the presence of Pluto. The sequence is: We *see the idea* and the possibility, we *feel* or relate emotionally to the change, we *experience* the transformation itself. The outer planets tend to add more intensity as well as expand the arena of experience.

Where do Venus and her higher octave, Persephone, fit in? Venus, ruler of Taurus, represents the body (bio-chemical) experience, as well as the first stages of social interaction via the senses. Persephone is the higher octave of this plane. Once we have thought, felt, and experienced transformation, we want to share it, to recycle the energy back into the world. When we *share* the experience of transformation, Persephone is present. She represents a more complex, more powerful, more conscious form of relating.

Astrological keywords traditionally associated with Libra such as diplomacy, law, peace, balance, marriage, are not words that fit the myths of Venus. On the contrary, her myths reveal her as much less conscious and primarily interested in her own personal pleasure as motivation, an indication that she rules Taurus. Venus may give us the sensual pleasures of relating; but Persephone reveals the power of relating, especially on a more conscious level.

It really should come as no surprise that the myths reveal Venus-Aphrodite as estranged from and in conflict with Demeter-Ceres and her daughter. The studies of female psychology, anthropology, and classical mythology all agree that there is a great disjunction of female sexuality-sensuousness and maternity-motherliness. There seems to be a universal conflict between the biological forces that unify these roles and the social ones that keep them discrete. A survey of Western culture will reveal that there is a strong tendency in myth, literature, religion, law, and implicit social *mores* to disassociate the erotic function of women from the maternal function. We only need to mention the invention of nursing bras and plastic pacifiers to illustrate the suppression of the sensuous aspect of mothering, the maiming of instinct, and the denial of the union of these two complexes.

Biologists have recognized and researched the physiological behaviors that are shared by undrugged, natural childbirth and the sexual excitement of coitus, as well as connections with lactation (the release of the hormone oxytocin during love-making, childbirth, and nursing is but one example). The behavioral analogies between sexual orgasm and "birth orgasm" are a striking embodiment of the essential unity of sensuality (Venus) and maternity (Demeter-Persephone). So why is there such an antithesis between erotic and maternal love in myth, religion and society? One explanation is that sex and sensuousness is associated with wildness, lack of control, a more primitive and less civilized order of being, while mothering is the initial model of social and civilized relationship. It is interesting to note that Demeter (the grain) was considered the goddess responsible for civilizing humanity via cooperation with nature and agriculture, while the god Dionysos (the vine) was more of an untamed entity.

On the other hand, motherhood has also been classed as being closer to the earth and nature, and correspondingly removed from reasoning, rationalizing, and whatever intellectual basis underlies the civilizing process. Most cultures have developed morality structures that rely on the distancing of sexuality-wildness from the civilized ideal of family-motherhood, often with the purpose of assuring a controlled economy with paternity-determined property rights. Moral attitudes can be severe; thus it is today that a mother nursing openly in public (and visibly enjoying it) is considered almost pornographic and suggestively incestuous. According to the laws of most cultures, the sin which is most nearly equivalent to the evil of murder or rape is *incest*.

Venus-Aphrodite and Demeter-Persephone epitomize two integrated sides to the feminine nature which have been deliberately separated. To allow them to be joined would imply a concentration of power that would be threatening to the male image of authority and the patriarchal dominance of mechanistic-materialism. The antithesis is another example of "keeping women in their place," but the projection of the Venus-Demeter split is not limited to women. Many women dichotomize men in the same way: the sensuous, artistic (and societally irresponsible) lover versus the solid and dependable family-man (who probably also fits society's economic ideal, buying the capitalist party line).

This antithesis is passed on in the emotional development of the child. Babies are initially open to the natural mix of heterosexual love and parent-child affection, but through training they learn to keep the two apart. Language cues are a key: we distinguish certain words and intonations that are used in courtship and adult love from the ones we use in parent-child relations. We are so unconscious about these differences that the subject is hardly even recognized in the fields of socio- and psycho-linguistics. There is quite a comment on human relationships to be found in the observation of the current vast divide between the archetypes of female lover and nurturing mother, and the male lover versus loving father.

This was not always so. There is evidence from early Mayan and Indo-European imagery that there was once an acceptable (or at least recognized) synthesis of sensuality and maternity. We might hope for a future that would re-synthesize the deep and natural connection of these divided concepts. It is simply a recognition and acceptance of our concrete lives as incorporating the more awesome and instinctual powers beyond our control.

THE MOTHER-DAUGHTER PAIR AND THE SON-LOVER

We have seen how Persephone has been astrologically mis-identified with both Pluto and Venus. There has also been confusion with the figure of her mother, astrologically symbolized by the asteroid Ceres and the sign Virgo (the Maiden). The figure of the constellation Virgo, ruled by Ceres, is a woman holding a seed or bud. The star that is in the figure's hand is Spica, which is the spike or bud of wheat or ear of corn, eternally reborn at harvest time. She is also sometimes pictured as holding a child. Persephone's son Dionysos is sometimes considered astrologically to be situated at Spica (currently located at 23 Libra tropical, 29 Virgo sidereal), a fixed star in the constellation Virgo, immediately adjacent to Libra (which is sometimes referred to as the Chariot with which Pluto abducted Persephone). Spica is one of the five brightest stars in the Zodiac belt, and was a major determinant for the ancients in constructing their calendars and astronomical systems. The ancient Egyptians and Greeks built numerous temples dedicated to goddesses that were aligned and oriented toward Spica.[52] With the advent of the patriarchy, however, many of the wholeness symbols were disguised and renamed. With the coming of Christianity, a Zodiac bearing a goddess with a special child (other than Mary with Jesus) could stand for heresy.

Dionysos was indeed a heretical god, the child of both light and darkness, the product of the union of the god of the Underworld and the goddess who treads the border of both worlds. Dionysos is the union of dualities. His message to mankind was the vine, the spiraling serpent power of Kundalini. He said, "Don't drink blood (kill), drink wine." In other words, "Make love, not war." The effect of wine or other intoxicants in celebration and ritual is to reduce the usually strong bastion of fears and inhibitions, a gentle way of discovering our selves within those whom we once believed to be our enemies or "separate" from us. The Piscean influence (the orientation toward dissolving boundaries to achieve union through being "under the influence") is clearly observed here, and numerous researchers have noted the similarities of Dionysos with the Piscean prototype Jesus Christ.

In Christian mythology, the Father and the Son are one. This Father-Son complex is characteristic of religions arising during the Precessional Age of Aries. In ancient matriarchal tradition, the Mother and Daughter are one. This *matrix* is typical of most religions dating from the Age of Taurus. The image of the Virgin Mary with the son Jesus (Persephone-Ceres-Virgo Maiden-Mother with the son Dionysos) is apparently the attempt of the Piscean Age to blend these two traditions, but the downside of Pisces is confusion and lack of discrimination. As we come into the Age of Aquarius, there may be greater clarity about this "mixed message." Mythology that evolves during the coming Aquarian Age is likely to take on the Aquarian quality of objectivity, and Persephone's simultaneous ingress into Virgo around the year 2011 (one Precessional Age after the birth of Jesus) may be a sign that we will become more able to discriminate and distinguish her identity.

Adding to the confusion (or richness) of images associated with Persephone, Demeter, Dionysos, Pluto, and Venus are the apparently similar figures from the myths from ancient Mesopotamia and Egypt. Diodorus Siculus mentions that *Isis* of Egypt was much the same deity as Demeter in Greece. Myths about the supreme Sumerian goddess *Inanna* have also been recognized as forerunners of the Greek Persephone myth. The Akkadian story of *Ishtar's Descent* is another story with similar features. All these goddesses were identified astronomically by the ancients as the planet Venus.[53] These and related myths tell of a descent by a female into the Underworld, with a male lover, brother, or son playing an important role in the story.[54] There is much we can learn from these related versions. Perhaps one of the most insightful portions of the Inanna myth is her initial response of being puzzled by the Underworld. Inanna asks repeatedly,"What is this?" And she is repeatedly told, "Quiet, Inanna, *the ways of the underworld are perfect. They may not be questioned.*"[55]

Around the ancient world the passage from winter to spring was observed by rituals and celebrated in legends of a promised return from darkness. Often these stories related the possibility of disaster from too rapid a search for ascent and resurrection. The warning was clear: one must fully experience the death or the depth of the descent, or else the returning process will be delayed further. All of the ancient rituals involved initiation ceremonies, processions, dancing, and other overt activities, but everyone knew there was a deeper meaning. This mystery of the first sign of spring was symbolic of the step toward a richer life in the development of the human soul.

We moderns think we are more progressed than the ancients, and that since science has replaced religion, we have no need for primitive ceremony. The modern world often ignores the understanding of individual development and the deep inner need for discovering that life advances by regular stages which are so important that they should be ritualized. Astrology serves a vital purpose because it recognizes and identifies these stages.

Chapter 6

NEW PLANET ON THE HORIZON[56]

Astrologers often use the axiom "As Above, So Below" to refer to the workings of the great Law of Analogy, sometimes called the Hermes Law of Correspondences, or the Jungian term Synchronicity. In plain language, it means that whatever is "up there" in the sky is also happening "down here" on earth, literally as well as symbolically. We can also extend this to mean that whatever is "out there" in the physical world is also "in here" within our consciousness. Johannes Kepler, last of the great astrologer/astronomers explained it this way, in his *De Stella Nova:*

"Nothing exists or happens in the visible heavens the significance of which is not extended further, by way of some occult principle, to the earth and the faculties of the natural things; and thus these animal faculties are affected here on earth exactly as the heavens themselves are affected."

It is easy enough to say that we understand this principle as an explanation for "why astrology works." Where the challenge comes in is to ponder the full implication, using the birthchart as a mirror of life by which we reflect and gain greater awareness. We accept the concept of mirroring[57] by interpreting the Ascendant-Descendant horizon in the following way: "As I am expressing myself as an individual through the Ascendant, so will there be a reflection back to me from the "Other" via the Descendant. Put less abstractly, *what you put out is what you get back.*

The horizon (Ascendant-Descendant) of the birthchart gives a key to understanding how we are all truly connected. As with the Self (the Ascendant), so it is with the Other (the Descendant). As with me, so with you. We experience individuality, our separate consciousness, via the Ascendant. We experience relationship through the Descendant. One is separation, the other union, yet together they comprise one continuous horizon.

The Ascendant is in the East, where the Sun (our waking consciousness) rises. The Descendant is in the West, where the Solar consciousness "dies" as it sets in the evening. Cultures throughout the world have in their mythologies the story of the Sun's cycle of day and night. It is a journey from birth to death, a representation of the transformation from self to not-self. What we have done in Western culture is to separate these two experiences.

Much of our experience in the mundane world defines life for us as polarities, as separate and opposing energies. Much of the stress we feel is the result of our identifying with the separated polarities and losing track of the "fact" of the continuity of the horizon. We get caught up in distractions and forget that Life is unified. Then we have moments of remembering. This back-and-forth of forgetting (separating) and remembering (reuniting) is the ongoing dance of consciousness.

One of the most highly exalted polarities in the world today is the separation into masculine and feminine, mind and feelings. An example of this is the situation of men fearing to express their feelings and women fearing to express their mental capacities. We justify our fears because there are consequences of socialized stress in crossing gender-identified borders. In other words, there is a great deal of social pressure to act in accordance with society's spoken and unspoken rules for male or female behavior. After thousands of years of cultural exaltation of this polarization, particularly with preference given to the patriarchy and the mind, there is now a movement to balance by emphasizing the so-called "Rebirth of the Feminine." But current experimentation with more androgynous modes of being threatens the secure familiarity of established gender-role definitions. As stated previously, the transition is not going to be gentle.

What we are doing is more than trying to meet the challenge of redefining how to be "separate but equal." We are accepting (remembering what we have forgotten) that there is a whole

59

continuum, the horizon, that incorporates the needs of both our Ascendant and Descendant. We look at our own lives as well as the broader goings-on of the world, and we recognize a dramatic crisis-in-consciousness evolving along the horizon-axis.

Astrologers have attempted to identify the astrological signature of this development, wondering if the crisis is fully accounted for by the Precessional transition into the Aquarian Age and the consequent resolution of Leo-individuality within the Aquarian group-unity-consciousness. Is it represented by major alignments of the known outer planets, or is something else afoot? As we described earlier, the pace and intensity of current earth changes hint strongly at the possibility that a new planet is once again in the process of being discovered, and the story of Persephone, with its themes of separation-reunion, seems ripe to pick.

TELLING THE STORY — AGAIN

Let us return once again to Persephone's myth to see what more can be revealed, especially with regard to our experience of the horizon. When it is time for Persephone to conclude her identity as Kore (youthful, innocent maiden), she is no longer protected by the Earth Mother force. In the original myth, it is Gaia, the Earth Mother who is older even than Demeter, who splits open the ground to allow Pluto entry from the Underworld. This tells us that there is permission granted by the feminine lineage, the inner knowing, to allow the maturation process to follow its natural, if violent, course. Persephone is captured and taken down under the surface of life and made Queen of the Underworld. Various reactions occur back "on earth," and through a series of adventures and arrangements, our heroine eventually enters into a perpetual pattern of separation and return to and from the waking world.

Persephone dies and is reborn in a constant cycle between worlds of earth-consciousness and subterranean consciousness, but whenever she returns to earth, she comes not alone. She has eaten the pomegranate seeds offered by Pluto (symbols of fertility and carnal knowledge), and now carries the treasure of deeper knowing back to earth. In many versions of the myth, she brings back a son Dionysos, child of the dance of consciousness. She has passed from innocent maiden-daughter to wife-Queen and mother herself. She represents the budding energy within that carries the seed of renewal and rejuvenation of life on earth.

Although she returns every spring to fulfill her earthly duties among the other sister goddesses, she also goes back down every autumn to join Pluto as consort and co-ruler of the Underworld, performing a sort of "double-duty" on the night-shift. Persephone is herself part of a separation, being one expression of the Mother-Daughter Matrix or the Goddess Trinity of Daughter-Mother-Crone. Persephone is one aspect of the Heroine's Journey through innocence-maturity-wisdom.

One of the paradoxes of being a woman which is depicted in the double Goddess of Demeter-Persephone is that motherhood is a kind of preparation for maidenhood. How is this so? Because when a woman has been split in two (divided) by being pregnant and birthing a child, she can know what it is to be whole. There is something about the experience of mothering that opens the way for women to experience individuality; the division of a pregnant woman into mother and child makes a woman more conscious of the fact that she can be an individual. Prior to that, her natural, unconscious ability is in relating. Likewise, men are naturally, but unconsciously individual. One way to look at this is to observe that men seem to start with knowing themselves; individualization seems to be more of a natural, if unconscious, state for them. In a way, men begin by knowing what it is to be whole; they have to "work" at relationship in the sense of what comes naturally for women, namely being aware and sensitive to another's needs. There is the opportunity in the androgynous model symbolized by the Pluto-Persephone myth for both men and women to become more conscious of both modes of being, individualizing and relating.

Persephone represents the eternal cycle of separation-reunion. She embodies the *crisis of separation* that is present in all relationships — with mother, with partners, with the earth itself. As an astrological planet, she would "rule" such issues as the transition into womanhood at the time of marriage and/or the beginning of the sex life, and all the variants on the basic theme of the emotional struggle of the feminine maturation process. Persephone has been negatively associated with the experience of smothering relationships where there is a fear of breaking away, as in the case of a possessive mother or divorce. In the positive sense, she is the diplomat connecting the worlds of light and dark for the purpose of maintaining balance and wholeness.

TRAVERSING THE BORDERS OF THE UNKNOWN

If we start at the center of our solar system and move from each planet to one further out, we see that each new planet opens up new vistas on human development, but each in turn also presents challenges or veils that hide the next planet out. Since Pluto's discovery in 1930, the earth and its inhabitants have had to experience what was previously unknown, the Scorpionic-8th house realm, but in a manner totally devoid of the feminine aspect which had been relegated by a patriarchal culture to the subconscious Underworld (in Persephone's myth, her abduction was planned with the collusion of Zeus-Jupiter, the male head authority of the Greek Pantheon). Much of what astrologers have previously associated with Pluto is actually ruled by Persephone, but has been veiled by Pluto. What is unknown has been labelled the "dark side of the force," and thrown out beyond Pluto, projected past the limits of a "known solar system."

For enlightenment purposes we can use the astrological solar system as a model for looking at ourselves. We can imagine ourselves standing on one planet and from that perspective view another planet. As an example, we can get into a Saturnian mode and take a look at our Uranian urges from a more disciplined standpoint. Or from Persephone we can get insights on Pluto. The fascinating experience with "standing" on Persephone is that we can both look inward at all the defined aspects of ourselves, but also look out upon the unknown. She is quite literally our link to outer space and that which exists beyond our known Self.

What an analogy is the solar system! What we know of ourselves and of life is symbolized by the known planets, while we shove what is unknown, uncontrollable, and fearful to us down into the subconscious, the Underworld (in our analogy, past the edge of the known solar system). As Pluto surfaces, he lets us know more about the experience of fear and death, and tells us that it is in order to transform. But Persephone reminds us that death is not a final loss, it is simply another form of separation. Since separation is only one part of an eternal cycle, nothing is ever really lost. (This is, by the way, a real clue to dealing with loss-lack consciousness.) Pluto awakens us to *why* we transform; Persephone shows us *how* we can transform, and *where* —- along the horizon of the Ascendant-Descendant.

As consort and "death partner" with Pluto, Persephone gives us a more complete picture of how individual transformation can occur through the experience of "letting go" in relationship and death (the ultimate separation). We can, in any circumstance, substitute the word *separation* for the word *death* and achieve a great leap in consciousness. The term *separation partnership* may sound like an oxymoron, but that is precisely what a conscious relationship is about. The success of partnerships (marriage, lovers, parents, friends) may not depend so much on "quality time" together, but rather on *quality separation*. In astrological charts (individual birthcharts as well as composite and synastry), we can look to the interaction and aspects of Pluto with Persephone-Transpluto for information about successful conscious relating.

This Persephone-Pluto pairing represents a higher octave awareness level (and consequently a more challenging one) than the usual Mars-Venus combination that astrologers look for in analyzing love relationships. Venus and Mars function together as the attraction of opposites. Their success depends, in fact, on exaggerated differences and heterosexuality. But once the biochemical buzz wears off (usually after six months to a year of sexual interaction, approximately one Venus cycle), the *relationship statute-of-limitations* runs out. Typically, the people involved start wondering what else they have in common and whether they can be companions, not just lovers. This is a point at which the "C-word" (commitment) comes up, and there is a choice presented to those involved about the possibility of the *conscious management of a relationship*. Part of that management involves the resolution of issues around separation. This can be potentially confusing because voluntary separation seems contradictory to the attraction principle upon which Mars-Venus is based.

The lowest levels of the opposing signs of the Taurus-Scorpio axis can be expressed as: "I desire you, therefore I own you." Ownership in relationship translates as possessiveness and jealousy ... the infamous green-eyed monster that dares and taunts us as relating beings. It could be said that the astrological Venus-Mars pair represent what the socio-biologists recognize as our inherited mating behaviors "left over" from earlier evolutionary stages of mankind. While we acknowledge the survival function of these primitive urges and instinctual hormonal responses, we also strive, in our spiritual development, to develop conscious management of these urges. In other words, we would like to enjoy the benefits of the Venus-Mars levels, without being dragged or led around by our emotions and hormones. How do we do this dance?

First, we have to see the possibility, and we only see by stepping back to observe, or by turning toward the mirror... this means separating! This is the mechanism implied by the ancient split that occurred within one sign Scorpio (representing instinctual sexuality, raw and deep emotion). As the Claws of the Scorpion transformed into the Scales of Libra, we divided ourselves as an evolutionary step forward. This is an analogy for our ability to lift our awareness center from being engulfed in an apparent wholeness that is our more emotional and unconscious self. This is an act of separation, to become objective and more conscious about relationship. It's not the easy path, however, because we have to let go of a lot of old, deep "stuff" that is anchored due to its purpose called *survival of the species.* And letting go of old baggage gives the apparent experience of "suffering." As any Libran will tell you, "It depends on how you look at it." We see first, *then* the integration happens.[58] We divide or separate, and then reunion happens again.

It is only through the process of repeated separation and reunion, the forgetting and remembering, the cycling dance of suffering and enlightenment, that our consciousness grows. Achieving higher awareness is not a static event or end result as pictured in the emphasis that is sometimes given to Persephone's "final" return to the mother. We can't hide there forever in the secure and static world of Mom and instinctual Nature. Evolving comes out of the ongoing process of alternating between aloneness-individualization and togetherness-wholeness. We are drawn toward experiences at both ends of the spectrum (horizon). All of us (women and men) take the Heroine's Journey. We individualize by separating from the mother, we face Pluto and death, we merge with him (become his bride), gaining our own kind of rulership. We remember (return to the conscious world), and then, oops—-it's forgetting time again, and back we go into an unconscious state for more "suffering," then reawakening again. The myth of Persephone shows us HOW consciousness is an ongoing, developed process.

There is no judgement placed on which end of the spectrum is better or worse (being more enlightened or conscious vs. "at effect" of victim-unconscious). It is not, strictly speaking, accurate to speak of an enlightened person, but rather of enlightened activity. The trick (or the mystery) is to use or manage separation as a technique. The trap lies in "getting caught up in" the separation itself. As often as we identify with our separateness, that many times do we have the opportunity to also re-unify. When we forget the essential unity of life, we end up stationing ourselves *here* and placing others "over there," blaming and finding mistakes in *them,* making them the victimizers and us the victims. To polarize in such as way is asking for struggle and confusion along the axis of the horizon. We forget the mirroring effect of the Descendant, we forget that what we see in others is also in ourselves. It doesn't work to continue isolating the egos (Aries) and attempt a makeshift compromise solution (Libra) by negotiating from a "stuck" standpoint across an opposition of polarities. The downside of the Libra is to see life as polarized opposites without a sense of the continuity of the horizon.

There is a way to accept diversity within the unity of all life. Persephone shows us how to move more freely along, above, and under the horizon, between the worlds of our collective and individual selves. The "I and Thou" of the Ascendant-Descendant are, in reality, just a larger whole called US.

UNIVERSAL THEMES

The theme of the Persephone myth is a universal one of the transition from an unconscious, youthful state to that of the blossoming maturity of awareness of the duality that exists within unity. The Persephone journey shares certain elements with other similar myths of Quest. There is the Descent away from the sheltered and familiar life, the encountering of perils and Danger along the way, the Overcoming of fears and inhibitions through meeting the challenging and deep experience. Ultimately the quest is for union with the soul or higher self, a connection with all parts of the consciousness. The quest is often objectified in myths as a treasure or magic amulet that will give the Hero-Heroine some unusual power, usually the ability to overthrow some manifestation of darkness (evil). This dark source (often appearing as a dragon or monster who abducts the maiden) exemplifies the Hero's own fears of death and change, and inhibitions about loss of control.

Many ancient stories, especially the Creation myths, tell of an initial state of wholeness or unity and innocence which must be shattered in order to make it possible to create the world as it is and the human race. The plot goes something like this: Once there was a sense of oneness, then a fracturing occurred, and the idea was created that god was separate from "man." As God ascended into the heavens, a counterbalancing event happened, namely that the Goddess descended into the Earth. In the history of civilization, as the patriarchy established itself as the ruling paradigm, anything which spoke to this original wholeness threatened the status quo and was deemed taboo. The cult of goddess-nature worship was forced to "go underground." Here was the Mystic Dissociation which explains the birth of the Occult Sciences. Astrology, Tarot, Runes, Kabala, all of the occult practices became children of the Goddess. The higher that god went into the sky, the more elevated did Mind and Science become. The more that "man" separated from Nature, the more were astrology and the occult associated with depths of darkness. The principle of separation which we know today as repression of the "dark" feminine (including persecution of witches, gays, astrology) is merely the converse of the exaltation of the masculine.

This idea that man was not the incarnation of divine life, but of a nature separate from god, has been essential to humanity's creating our "reality" as it is today. What was the purpose of exalting mind over matter and spirit over the soul? We look around and see that we have gained mental and technological dominion over the physical earth. Repression of the Goddess was a functional and necessary polarization that allowed us to expand our minds (the masculine principle) and manipulate the physical world (the feminine principle).

The trouble with this approach is when we forget to apply the principle inwardly, so it manifests as an outwardly manipulative domination over Nature. Thus have we set ourselves up for eventual rebellion, which is precisely what is happening today. The world has become polluted and imbalanced on all four elemental levels.[59] The FIRE element is polluted as the random mutation of atomic power. The EARTH element is polluted as materialism gone rampant and garbage we can't deal with. Our atmosphere and oceans (the AIR and WATER envelope of the planet) resemble bungled chemistry experiments.

What about us? Humanity's consciousness is polluted with fear-based and hormone-driven attitudes. We have reached a limit in identifying "progress" as an attempt to control Nature with the mind. (Rob Hand has called progress the "heaven-on-earth" of the modern religion of Scientism, or Mechanist-Materialism)[60] Because the emphasis has been on Mind-over-matter, that which is below the conscious mind is pushing to be born. We labor as if with birth contractions to let what is below consciousness arise into the realm of the waking self. Fears become unmasked, and we discover it is simply new energy released through the experience of unity. We can see this happening as we attempt to reunite Astrology and all other aspects of the Occult with the light of day and scientific knowledge.

REVIEWING THE MYTH

Let us go back over Persephone's myth again. In the Greco-Roman version she is the daughter of Jupiter and Ceres (Zeus and Demeter). Her mother adored her and protected her jealously. One day, Pluto saw Persephone, and prompted by the arrow of Eros, fell in love, asking his brother Jupiter for her hand. Jupiter liked this idea, but knew Ceres would not give up her daughter easily. That which Ceres feared, separation, became manifest. Pluto and Jupiter plotted Persephone's abduction and gained the consent and cooperation of the Greater Mother Goddess, Gaia.

Persephone was playing in a meadow with her nymph friends, picking wild flowers, the image of youthful innocence. Pluto and Jupiter had persuaded Gaia to make a special flower to grow among the others, the hundred-petaled Narcissus. Persephone saw it, was struck with its beauty, and picked it. She chose it, she wanted it, she uprooted it, separating it from the earth. Persephone chose to know herself (via the Narcissus, flower of self-consciousness), and the earth opened up (by permission of Gaia). We have here a situation in which permission has been granted by both the self and feminine destiny. This is not a fateful act instigated by external motives; the "rape" was not a violation that happened to Persephone against her will.

As Gaia allowed the earth to split open, Pluto flew out in a blaze of glory, riding his chariot of flames. He grabbed the maiden and bore her home to Hades, while the earth closed up, sweeping in the swine that were grazing nearby. (We shall hear much more about these swine in later chapters. Let it be known for now that pigs were the sacred animals of both Persephone and Demeter, which were sacrificed before the planting of new corn, their carcasses being thrown down into pits in the earth. We note also in passing that the patriarchal Hebrews forbade consuming pig meat and believed them to be vile animals. This is perhaps because pigs were too familiar as indicators of Goddess-worship and therefore had to be repressed and forbidden. Such was the case with the wrath of Moses upon descending from the mountain to find people worshipping another Goddess animal, the golden calf.)

When Pluto took Persephone to Hades, he made her his Bride and gave her the title Judge of the Dead. There is no association here with the Christian idea of judging the dead for deeds in their recent life. The "dead" translates here as unconscious or unmanifested potential (buried treasure). Pluto guards the dead and will not allow the unconscious treasures or secrets to become conscious except through union with Persephone. She is the one who carries these secrets into the light world. When we reach for power out of fear, that is Pluto's expression, and death is the result, created through random growth such as obsessive power gone wild, suicide, AIDS. Persephone directs such power and growth through maintaining balance between dualities. She represents the conscious feminine essence. Persephone's "judgement" is reflection. She turns the mirror toward us, and if harmony is perceived, then the soul passes to another level of existence. If disharmony is seen reflected, the soul returns for rebirth or repeat of the experience.

In the Tarot, she is the dancer of the world-universe card (#21). Those who have written of the meaning of Tarot cards have said this card indicates "dancing on your limitations,"[61] "the cosmic communion of souls,"[62] "the freedom and power to step out and make a choice,"[63] and the possibility "to see things as they really are, ... in harmony with the universe."[64] These descriptions seem in accord with Persephone's myth.

Meanwhile (back to the myth), Ceres searches everywhere for her daughter, and this searching yields many wonderful subplots. When she finally finds out about the deception, she is so angry that she stops plants from growing. So the earth experiences its first winter as a result of the divine mother's exceptional grief at the loss of her beloved daughter. Conditions become intolerable on earth, people are starving. The gods on Olympus begin to worry about whether mankind will perish (The myth explains that their worry is not about the survival of humanity per se, but a concern that the gods would, in turn, perish for lack of someone to worship them!).

After much negotiation, Pluto finally agrees to release his new bride. The very first thing that Ceres says to her daughter upon meeting her is not a greeting nor expression of relief, but the anxious question, "Did you eat anything while you were in Pluto's house?" Persephone admits that she has eaten pomegranate seeds, and thus is her fate sealed as wife to Pluto. The symbolism of eating the seed is usually interpreted as the awakening of sexual awareness and/or becoming impregnated. The myths vary as to how many seeds she consumes; usually, it is one or three.

The conclusion of the myth includes the arrangement for Persephone to spend part of the year with her husband, and part with her mother, presumably performing goddess nature chores. The total time of her visits to both worlds is, in every version of the myth, always dependent upon the number of seeds she has eaten. With Ceres' joy upon the return of her daughter comes a magnificent spring and a bountiful crop. The reflection of this on the daughter's part is to return bearing a child, Dionysos. Later on comes the first autumn as Persephone readies to rejoin Pluto and Ceres becomes sad (the earth mourns), even though she knows that Persephone will return again the following spring.

One factor which we must remember is that Persephone as a mythological character is not isolated. Her story is not simply "the Persephone myth," or even "the Demeter-Persephone myth." It is also the *Pluto-Persephone myth.* Although this book attempts to cover the subject of Persephone as completely as possible, we are (in a sense) only telling half the story. What about the Pluto half of this tale? We cannot assume that Pluto is in Hades all the time. Perhaps his capture of Persephone had some further purpose.

We know from some sources that Persephone's position as Queen of the Underworld fulfilled a need. It is said in the oldest version of the myth that, before Persephone arrived, Hades was a place of nothingness and that souls there were lost in a sort of limbo. The story goes that when Persephone heard the mournful cries from the Underworld, she felt urged to be of assistance, descended of her own volition, and came to the lost souls as the sign of hope and renewal. But meanwhile, what was Pluto doing? Did he want Persephone to be Queen in order to help "run" the Underworld while he was busy with his other tasks? Persephone has a dual purpose, to assist her mother on earth and to be Queen of Hades. It may be that Pluto plays a counterpart role to Persephone's; that he also has "another life." Such a concept may explain how the Earth-Underworld inter-change, or the conscious-subconscious dialogue, may have different manifestations for men and women. Does Pluto have a relationship to Mars as Persephone does with Venus? These are questions that might be addressed in a closer examination of the myths of Pluto, a topic which could fill another book.

Chapter 7

THE MYTHOLOGY OF SCIENTIFIC DISCOVERY

Historians of science like to relate the stories of an invention or discovery as a final triumph of the Scientific Method, as though it were the final destination along a linear path that has been carefully prepared by orderly, logical thought. But real Science just doesn't happen in such an ideal and rational way. The "path" of discovery is usually more of a meandering, with deadends, backtracking, criss-crossing, and a jumbled fusion of input from several directions of thought, not all of which are traditionally "scientific" or even morally "pure" (many scientific "first ideas" have been stolen or borrowed or connected with massaging someone's ego). It is easy enough to go backwards from any event and restructure history so it looks like a train of responsible developments leading in a tidy and consistent fashion to the event itself. (Astrologers, by the way, can be as adept as scientists at performing this kind of revisionism.)

What is often left out of the story of a scientific discovery are the years of bungling trial-and-error, as well as the inspired moments in which imagination outweighed reason but happened not to produce "tangible" results in conformity with the final, officially sanctioned, and biased version of the discovery.[65] Such non-relevant brainstorming is often later redefined as the "crank" hypothesis, and therefore omitted from "serious" scientific discussion and the orthodox textbooks that teach history of science ... unless, of course, the "failed" prediction has been given by "one of their own." An example of this is the astronomer William H. Pickering, who (along with Percival Lowell) helped lead the way to the discovery of Pluto, and who predicted no fewer than eleven hypothetical planets, one of which he determined must be located at an astounding distance of 6250 a.u., with a period of half a million years.[66] Notice, however, the language used in the history of science in treating this outrageous prediction: "All in all, Pickering's

71

planets are a strange and various lot, remembered today (if they are remembered at all) as astronomical curiosities."[67] The scientific bias is clear: if an astronomer predicts wildly, and it's a scientific curiosity... Anyone else predicts, and it's a crank hypothesis. Such inherent prejudice is due to what the historian Morris Berman calls the Myth of Objective Consciousness.

Science has devised an entire personality that it projects to the public, attempting to impress an image of the professional scientist using his exclusively scientific methods to make the grand inventions and discoveries. This image stands in direct conflict with facts about which Science seems to be completely "in denial." One of these realities is the role of amateurs in planetary discovery. Uranus and Pluto were discovered by amateur-hobby astronomers; Pallas, Vesta, and countless other asteroids were, too.

The fact is that the discovery of a planet is never an isolated, self-contained event. But when scientific mythologists present the story, much of the background is either never mentioned or else is dismissed in the superior attitude so typical of omniscient Science. As an example, we quote here the introduction to one official rendition of "The Discovery of the Planet Neptune:"[68]

"Astronomers receive letters from time to time telling of remarkable astronomical discoveries made by the writer... Usually, such a letter receives a courteous, non-committal reply, is endorsed "crank," and then filed with others of its kind and forgotten... Galle, chief assistant at the Berlin Observatory, had read several communications of this type before receiving the now famous letter... telling him to turn his telescope upon a certain spot in the heavens [in order to] find a planet hitherto unknown...This letter did not go into Dr. Galle's 'crank file,'... for the writer was Joseph Leverrier"... [and here follows the author's list of Leverrier's scientific achievements which made him one of the Elect and therefore someone worth listening to].[69]

The author goes on to describe only the pertinent events which "belong" to the Scientific Establishment's official legend, a chapter in what we might call the Mythology of Science. Instead of describing the adventures of gods or universal heroes on a Quest, the Mythology of Science tells of competition among scientific professionals manipulating formulas and equipment. Anything of a non-scientific nature is treated like an evil dragon that must be conquered by Mind. In our exemplary scientific myth related above, even Bode's Law received a curt dismissal by the author describing the solution of the problem of locating a new planet: "Here [astronomers] were misled by a curious relationship among the planetary distances out to the orbit of Uranus, which is generally known as Bode's Law, though it is not a law in any physical sense at all, but merely a numerical oddity." In seeming contradiction to this remark, the author later says that Bode's Law actually fits the next planet beyond Neptune (Pluto) very well. He concludes that this "most notable episode in the history of astronomy," the discovery of a planet by calculation, "afforded the perfect proof of the validity of Newton's great law."

It is curious that scientists persist in calling Newton's Law "great," and Bode's Law a mere coincidence, especially in the light of what they should have learned from their own modern physics. The principles of Quantum Mechanics (which could be named the science of the twentieth century) in fact have the potential to dissolve many barriers erected by the fragmented world view of Establishment Science, as well as to support such astrological concepts as synchronicity. The old Newtonian world of separability claimed that things no longer in contact with each other cannot affect one another, but Quantum Mechanics violates this old principle. Despite the general ignorance around this issue within the established scientific community, there are some modern physicists who understand that consciousness modifies the quantum wave and thereby changes the physical universe. Here are their explanations:

"The recognition that physical objects and spiritual values have a very similar kind of reality... is the only known point of view which is consistent with Quantum Mechanics."[70]

"The doctrine that the world is made up of objects whose existence is independent of human consciousness turns out to be in conflict with quantum mechanics and with facts established by experiment."[71]

"The world cannot be analyzed into separate and independently existing parts... The sphere of ordinary material life and the sphere of mystical experience have a certain shared order."[72]

Despite the modern developments in physics, many scientists haven't taken such interpretations seriously. They don't realize the social and philosophical implications of these theories, but rather continue to support a world view that is based on a mechanistic (Newtonian) causality and separability.[73] They consider the parallels of quantum physics and ancient mysticism as an intellectual game. The elitism of such scientists who have bought or sold out to Establishment Science has alienated many people. "Real Science" is performed almost exclusively by professionals, who, like the authoritative Latin-speaking priests they replaced, communicate in a specialized language and brand outsiders as heretics practicing "pseudo-science."

SCIENTISTS DEBUNKING ASTROLOGY OR...
ASTROLOGERS DEBUNKING SCIENCE?

Popularity of the new alternatives to old Science is partly a reaction to the failure of the traditional scientific community to give meaning to life, and to the tendency of this establishment to exclude non-experts. Scientists often reject what they have labeled "pseudo-science" on the basis of the social characteristics of its supporters, while ignoring or failing to investigate the alternative theories themselves. Woe to any scientist who ventures beyond the borders to investigate the theories of these "fringe" areas, alternative studies or unorthodox disciplines; they are likely to be ostracized.[74]

It is an interesting commentary on the so-called freedom of speech that astrologers and other societal outcasts or scientific rejects are more open to coordinating with those who have cast them out than the reverse. It is yet another example of the Persephone theme of separation-reunion. The separation of Science from mysticism-intuition is nothing more than the old male vs. female mind-feelings split. The same fear is at work ... patriarchal fear of the dark feminine. "If I'm *not separate* (defined by my individuality), I might fall into the (w)hole. If I start to *feel*, I might lose my MIND."

Why do scientists have such a violent reaction to quasi-scientific systems like Astrology? It is more than just a difference of approach, for scientists make their judgments about astrology without seriously examining the theories. Scientists barricade themselves into a separate, protected turf, and they are scared and insecure, so they continue to polarize away from that which they fear joining. Thus we have persecution against the intuitive arts.

What it comes down to is an issue of power; scientists fear they will lose control of the power they believe they hold in intellectual society. Scientists feel threatened that people aren't learning science in the proscribed manner. The fact is, books and media presented on the occult, extraterrestrial life, and science fiction have done much to popularize the principles of modern science. Much of the modern public interest in cosmology (the scientific Creation Myth) has derived more from a fascination with the paranormal (the mysteries of life) than from a pure desire to learn how establishment science views the universe.

People aren't watching Star Trek just because it's more entertaining than a textbook. The "world" of Star Trek actually represents an accurate expression of the latest modern physics. The "transporter effect" and the "worm hole" are but two examples of how Star Trek presents the world as explained by quantum physics.[75]

Cutting-edge scientists and modern physics have presented us with concepts that support the re-association of Science with Mysticism and Astrology. They are "discovering" what astrologers have always "known." They have given it names like Quantum Mechanics, Gaia Hypothesis, Symbiosis, Morphic Resonance, and Grand Unification Theory, but to Astrology it has always been *As Above, So Below.*[76]

Scientists generally choose the worst kind of astrology (newspaper Sun-Sign columns) for their attacks. Astrologers could likewise point out examples of the worst kind of science as an offensive tactic. But we must get beyond such immature behavior which is based on mutual ignorance. Scientists must make serious examination of astrological theory, and astrologers must educate themselves about science. We must learn to distinguish the Scientific Method (which is perfectly valid) from the established Institution of Science, as well as from the prejudices of the Religion of Scientism.[77]

Astrology alone does not have all the answers, any more than does Science. But the two can complement each other. Astrologers must ask themselves if they are updating their methods or making the same mistake as some scientists, namely perpetuating outmoded Newtonian approaches of fragmentation (such as producing "cookbook astrology") and emphasizing the parts rather than the whole. Defining the birthchart as a unique moment, placed on a single point in space, along a linear construct of time, is in accordance with Newtonian "absolute time," which Quantum Physics has determined is obsolete. One way to incorporate the principles of modern physics into astrology would be to give greater weight to the *relationship* of aspects in a chart than to the individual planets involved, and pay more attention to broader *patterns of interconnection* than to exact, isolated transits to fixed points in a birthchart.

It is a worthless endeavor to argue for the superiority or greater validity of Science versus Astrology. It is worthwhile to open the mind to seeing the value of each, to make use of each in appropriate circumstances, and to encourage a more balanced relationship and integration of their two functions. This is what Science and Astrology can have in common. Joining forces might even bring about grander discoveries than each field can accomplish in isolation from the other.[78] An obvious example of possible collaboration is the search for Persephone-Transpluto.

Astrologers may be tempted to skip the next chapters because they contain so much scientific information. To do so would be to miss the entire point of this last discussion on Science and Astrology.

Chapter 8

THE RESONANCE BETWEEN ASTROLOGY
AND ASTRONOMY:
DISCOVERY OF THE ASTEROIDS
AND THE OUTER PLANETS

As stated in Chapter One, the discovery of Uranus introduced an entirely new approach to viewing the solar system. The planet Uranus was the evidence, the living "proof," that the solar system was not bounded by Saturn, the astrological planet of limitation. In his typical Uranian manner, he challenged the old tradition by shattering the established boundaries set by Saturn. He also presented the idea that planets were discoverable, and that therefore there might be others. The fact that Uranus was the first discovered planet means that astrologically we can use the chart of his initial sighting as a "birthchart" for the process of planetary discovery itself. We can compare his chart with those of the later-discovered planets, seeking correlations that might lead us eventually to conclusions about the nature of Persephone's discovery.

Looking at the discovery chart for Uranus (page 78), we notice that all the planets not yet discovered are involved in patterns of inter-aspect with each other. Persephone at 6 Aries is opposed Neptune at 4 Libra and sextile Pluto at 6 Aquarius. Ceres at 4 Pisces and Vesta at 7 Cancer make aspects to these points, as does Chiron (who is situated at his eventual discovery point, 3 Taurus). Juno makes a quintile with Pallas, who is trine Uranus, who is in turn quintile to Chiron. (All the aspects mentioned here are in very tight orb.) We note also the Moon's position conjunct the Ascendant in Scorpio, sign of revealed secrets, and located at one of the Gates of Power (15 degrees of the Fixed Signs). There are a fair number of corresponding inter-aspects in the discovery charts for Ceres, Neptune, Pluto, and Chiron. The most obvious are the following:

Persephone Is Transpluto

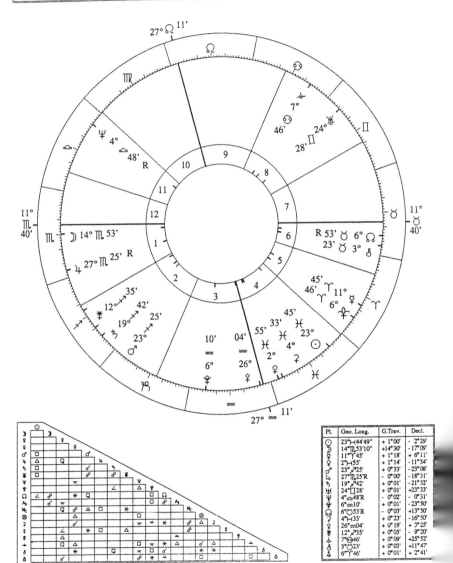

Discovery of Uranus	10:30:00 PM LMT	Mar. 13, 1781
Greenwich, ENG	51°N 29'00"	0°W 00'00"
Placidus	9:57:36 S.T.	J.D. = 2371629.4373
Natal Chart	R.A.M.S. = 23:27:36	Obl. = 23°28'04"
Tropical	No Ayanamsha	Geocentric Ecliptic

Pl.	Geo. Long.	G.Trav.	Decl.
☉	23°♈(44'49"	+ 1°00'	- 2°29'
☽	14°♏53'10"	+14°30'	-17°09'
☿	11°♈45'	+ 1°18'	+ 6°11'
♀	2°♓55'	+ 1°14'	-11°34'
♂	23°♐25'	+ 0°33'	-23°08'
♃	27°♏25'R	- 0°00'	-18°31'
♄	19°♐42'	+ 0°01'	-21°32'
♅	24°♊28'	+ 0°01'	+23°33'
♆	6°♒10'	- 0°02'	- 0°31'
♇	4°♉48'R	- 0°03'	+13°50'
☊	6°♉53'R	- 0°03'	+13°50'
⚷	26°≈04'	+ 0°19'	+ 3°25'
⚴	12°♐35'	+ 0°05'	- 9°20'
⚵	7°♊46'	+ 0°09'	+25°52'
⚶	3°♈23'	+ 0°03'	+11°47'
⚳	6°♈46'	+ 0°01'	+ 2°41'

Turning first to the chart of Ceres, the next planet discovered after Uranus (see chart, page 80), we notice that Pluto is at 3 Pisces, near the Venus-Ceres conjunction of the *Uranus chart*. Persephone is exactly stationary-direct and square her own position in the *Pluto discovery chart*. Ceres herself was discovered at 23 Taurus, within minutes of her station. The Moon is again in a water sign.

Neptune's discovery chart (page 81) shows a Persephone-Ceres conjunction at 4-7 degrees of Gemini, and Pallas is at 6 Aries, all aspecting their original positions in the *Uranus chart*. Saturn and Neptune are conjunct, sextile to Pluto, trine to Chiron, with all four aspecting the Uranus-Pallas trine of the *Uranus discovery chart*. The Moon is again in Scorpio.

Pluto's discovery chart (see page 82) shows a Ceres-Jupiter conjunction at 6-7 Gemini, Venus is at 3 Pisces, near the Venus-Ceres conjunction of the *Uranus discovery chart*. Persephone is aspecting the Uranus-Pallas trine of the *Uranus chart,* and the Moon is once again in Scorpio.

In Chiron's discovery chart (page 83), there is a stellium in Scorpio that includes Pallas, Ceres, and Uranus; a Pluto-Vesta-Venus-MC conjunction which is sextile to Persephone and square to the *Pluto discovery point*. Persephone, at 17 Leo 40, is within 6' of exact semi-sextile with *Pluto's discovery point*. The Moon is, as in the Ceres discovery chart, in the water sign Cancer.

DISCOVERIES THAT DIDN'T HAPPEN

The history of planetary discovery is, in fact, very much a history of missed opportunities. Planet-hunting has proven to be a most frustrating business. Among the astronomers who have deliberately set out to increase the size of the solar system, only a few have been successful, and even then their success has often been tainted by controversies over the nature of the discovery, over who claimed prior sighting, and over the scientific validity of their work that produced the discovery.[79] For astrologers, this means there is another area for research that might prove fertile for revealing the process of planetary

Persephone Is Transpluto

Discovery of Ceres	06:00:00 PM LMT	Jan. 1, 1801
Palermo, ITALY	38°N 08'00"	13°E 22'00"
Placidus	1:37:04 S.T.	J.D. = 2378862.2502
Natal Chart	R.A.M.S. = 18:43:36	Obl. = 23°27'55"
Tropical	No Ayanamsha	Geocentric Ecliptic

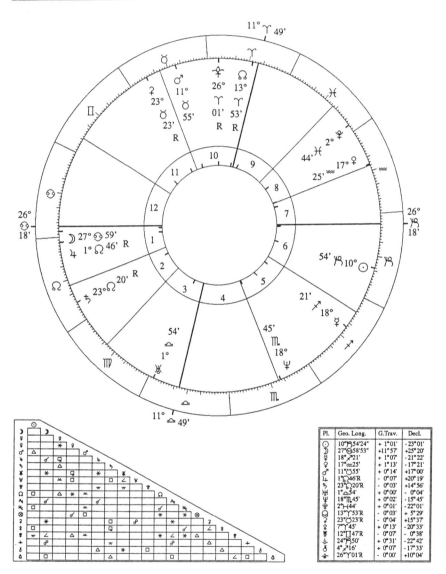

Discovery of Neptune	09:00:00 PM LMT	Sep. 23, 1846
Berlin, GER		13°E 25'00"
Placidus	21:02:47 S.T.	J.D. = 2395563.3334
Natal Chart	R.A.M.S. = 12:09:07	Obl. = 23°27'33"
Tropical	No Ayanamsha	Geocentric Ecliptic

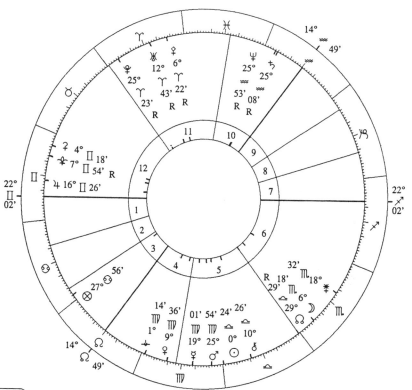

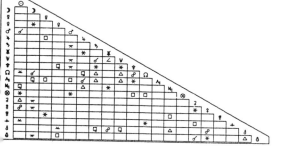

Pl.	Geo. Long.	G.Trav.	Decl.
☉	0°≏23'31"	+ 0°59'	- 0°09'
☽	6°♏18'02"	+12°33'	- 12°56'
☿	19°♍01'	+ 1°48'	+ 6°04'
♀	9°♍36'	+ 1°14'	+ 9°09'
♂	25°♍54'	+ 0°39'	+ 2°30'
♃	16°♊26'	+ 0°02'	+22°00'
♄	25°≈08'R	- 0°03'	- 14°40'
♅	12°♈43'R	- 0°02'	+ 4°22'
♆	25°≈53'R	- 0°01'	- 13°24'
☊	25°♈23'R	- 0°01'	- 6°22'
⊕	29°≏29'k	- 0°03'	- 11°18'
⚷	4°♊18'	+ 0°03'	+13°19'
♀	6°♈22'R	- 0°17'	- 8°29'
⚸	18°♏32'	+ 0°17'	- 7°13'
⚶	1°♍14'	+ 0°27'	+14°00'
⚵	10°≏26'	+ 0°09'	- 5°49'
⚳	7°♊54'R	- 0°00'	+21°39'

81

Discovery of Pluto	04:00:00 PM MST	Feb. 18, 1930
Flagstaff, AZ	35°N 11'53"	111°W 39'02"
Placidus	1:26:06 S.T.	J.D. = 2426026.4586
Natal Chart	R.A.M.S. = 21:52:43	Obl. = 23°26'54"
Tropical	No Ayanamsha	Geocentric Ecliptic

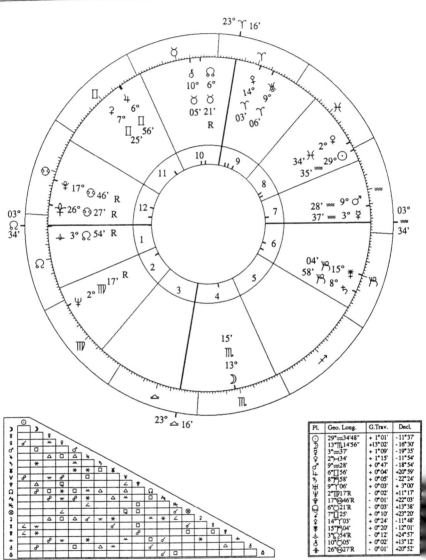

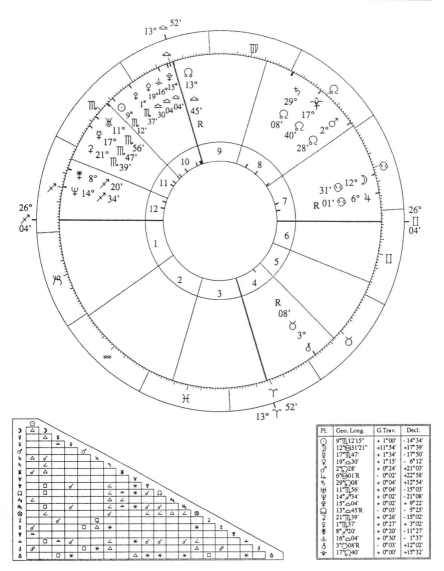

	Discovery of Chiron	10:00:00 AM PST	Nov. 1, 1977
Pasadena, CA	34°N 09'00"	118°W 09'00"
Placidus	12:51:02 S.T.	J.D. = 2443449.2506
Natal Chart	R.A.M.S. = 14:43:38	Obl. = 23°26'32"
Tropical	No Ayanamsha	Geocentric Ecliptic

Pl.	Geo. Long.	G.Trav.	Decl.
☉	9°♏12'15"	+ 1°00'	- 14°34'
☽	12°♋31'21"	+11°54'	+17°39'
☿	17°♏47'	+ 1°34'	- 17°50'
♀	19°♎30'	+ 1°15'	- 6°12'
♂	2°♌28'	+ 0°24'	+21°03'
♃	6°♋01'R	- 0°02'	+22°58'
♄	29°♌08'	+ 0°04'	+12°54'
♅	11°♏56'	+ 0°04'	- 15°03'
♆	14°♐34'	+ 0°02'	- 21°08'
♇	15°♎04'	+ 0°02'	+ 9°22'
☊	13°♎45'R	- 0°03'	- 5°25'
?	21°♏39'	+ 0°26'	- 15°02'
?	1°♏37'	+ 0°27'	+ 3°02'
⚷	8°♐20'	+ 0°20'	- 11°27'
⚵	16°♎04'	+ 0°30'	- 1°37'
⚳	3°♋08'R	- 0°03'	+12°02'
?	17°♌40'	+ 0°00'	+15°32'

discovery, the "close, but no cigar" category of "almost discoveries," the ones that just missed happening, were premature and too much ahead of their time to achieve fulfillment through official recognition.

All three of the major planets discovered in modern times were actually observed upon several occasions prior to the dates of their official "first" sightings. There are twenty-two pre-discovery observations of Uranus on record. The astronomer John Flamsteed saw Uranus in 1715, Joseph Laland recorded seeing Neptune in 1795 (May 8 and May 10, in Paris), and several photographs of Pluto were taken before 1930. In each case, the observer mistook the planet for a star, or thought he had been mistaken in observing the motion that distinguishes a planet from a star, and consequently the official discovery was delayed by many years.

Such *pre-discovery observations* are quite important to astronomers because they can be used to confirm or correct the calculations of orbit. Neptune and Pluto have not yet been observed during one complete orbit since their official discoveries; it is not unlikely that many corrections will be required in order to match new observations. Neptune does not return to its discovery point until about 2009; Pluto will take another 185 years to complete one cycle. Having confidence in the determination of known planetary orbits is a prerequisite for the accurate prediction of Persephone's orbit. It was, after all, the observed perturbations to the predicted orbits of Uranus, Neptune and Pluto that have led astronomers to seek for at least one Trans-plutonian planet.

It might be useful for readers to acquaint themselves with some of the terminology used by astronomers in planetary search. Five elements are necessary to define a planetary orbit. The *mean distance from the sun* and the *eccentricity* give the size and shape of the orbital ellipse. The elements that together give the orientation of the orbit in space and in relation to the sun are: *inclination* of the orbit to the ecliptic, the *longitude of the ascending node* (which is where the orbit intersects the ecliptic and which astrologers call the North Node), and the *longitude of the planet's perihelion*. To find a planet's position in such an orbit, two additional elements are needed: the *longitude of the planet* on some specific date, and the *mean daily or annual motion*.

There are a number of factors that complicate the determination of a planet's orbit and position. Each planet is irregularly perturbed from its course by the gravitational influences of its neighbors. Observations of a planet are compared with the theoretical orbit, and there are likely to be differences, called *residuals*. This problem is a form of the classic *three-body problem,* which permits only an approximate solution. Celestial mechanics is used to tinker with the orbital elements in order to achieve a better match between theory and observation. When such tinkering will not suffice, and when residuals remain larger than error can explain, it is assumed that unexplained perturbations are caused by some unknown gravitational factor (i.e., an unknown planet).

ELUSIVE AND CONTROVERSIAL NEPTUNE

The case of Neptune, astrological ruler of hidden, invisible Pisces, is perhaps one of the most interesting examples of unnoticed observation.[80] In his official discovery chart, Neptune was situated very close to conjunction with Saturn in Aquarius. There is potential for a new planet to be discovered in conjunction with a known planet because there is a likelihood that the observer is paying close attention to the known planet and then easily notices the relative movement of the nearby "wanderer." This is precisely what happened when Neptune was occulted by Jupiter during the winter of 1612-1613, and Galileo happened to be observing Jupiter.

Galileo's notebooks record his observations of Neptune from Dec. 28, 1612 through Jan. 28, 1613. His first observation, accompanied by an illustration of the moving body of Neptune, is dated Dec. 28, 1612 at 3:46 a.m. local time. (See chart below) In this chart, even the as-yet-undiscovered planets make aspects to their positions in the eventual "first" outer planet discovery chart of Uranus. Neptune aspects the Pallas-Uranus trine of the Uranus discovery chart, Uranus is half a degree from his discovery point. In addition, Pluto is at 3 Taurus (Chiron's discovery point). The Moon is in a water sign Pisces.

However, there are some revealing differences between the charts for Galileo's sighting and the later, officially recognized one. This latter chart features the discovered planet conjunct Saturn, the authority, which is situated in the sign of its traditional rulership Aquarius. The

Persephone Is Transpluto

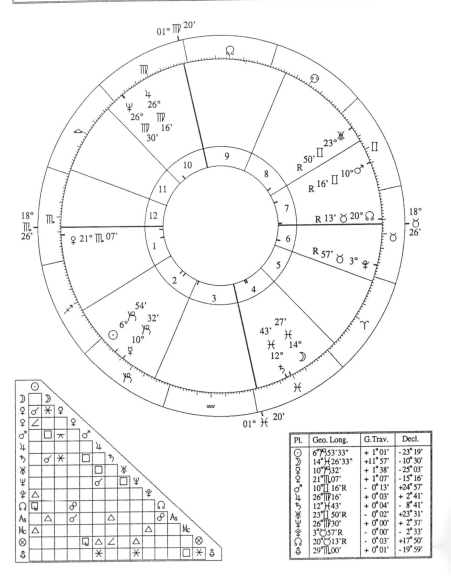

Sighting of Neptune	03:46:00 AM LMT	Dec. 28, 1612
Florence, ITALY	43°N 46'00"	11°E 15'00"
Placidus	10:58:37 S.T.	J.D. = 2310192.6577
Natal Chart	R.A.M.S. = 18:27:37	Obl. = 23°29'23"
Tropical	No Ayanamsha	Geocentric Ecliptic

Pl.	Geo. Long.	G.Trav.	Decl.
☉	6°♑53'33"	+ 1° 01'	- 23° 19'
☽	14°♓26'33"	+11° 57'	- 10° 30'
☿	10°♑32'	+ 1° 38'	- 25° 03'
♀	21°♏07'	+ 1° 07'	- 15° 16'
♂	10°♊16'R	- 0° 13'	+24° 57'
♃	26°♍16'	+ 0° 03'	+ 2° 41'
♄	12°♓43'	+ 0° 04'	- 8° 41'
♅	23°♊50'R	- 0° 02'	+23° 31'
♆	26°♍30'	+ 0° 00'	+ 2° 37'
♇	3°♉57'R	- 0° 00'	- 2° 33'
☊	20°♉13'R	- 0° 03'	+17° 50'
⊕	29°♏00'	+ 0° 01'	-19° 59'

86

ascendant is Gemini, with the chart ruler Mercury in the sign of his traditional rulership, and conjunct Mars. These aspects are reflected in the fact that Neptune was the planet "found on a piece of paper," anticipated by astronomer's logical calculations, and that there was at the time much dispute and verbal attacks over who could really claim the discovery, Adams or Leverrier.

The chart for Galileo's sighting, on the other hand, shows Neptune conjunct Jupiter in Virgo (a sign of detriment for both). The ascendant is Scorpio, with the then-known ruler Mars in a square to a Moon-Saturn conjunction in Pisces (Neptune's sign). Although Galileo tracked the motion of Neptune relative to Jupiter for a month, and would have understood that this was no normal background star, he made no claims of discovery. It is unknown why he did not announce his observation, but at that time he was already an open champion of the Copernican system, and by 1613 was involved in controversies that would eventually lead to his denunciation by the Catholic Church. Perhaps, to use a Neptunian analogy, he felt he was already in deep enough waters already without further rocking the boat. As with most Neptunian stories, this one remains a mystery.[81]

What is not a mystery is the implication of Galileo's observations. If his diagram is drawn to scale, as some astronomers believe,[82] then the presence of a tenth planet, one beyond Pluto, is almost a certainty. This is because the position of Neptune as indicated by Galileo's drawing does not conform to where Neptune should have been in his orbit, an anomaly which can be explained only by perturbation from an unknown planet.

The official discovery of Neptune coincided with the emergence of American astronomy on the world scientific scene and the entry of American astronomers and mathematicians into the history of planetary discovery. Harvard was at this time installing America's first large telescope, and the first issue of *Astronomical Journal* was published. In this regard, we make note of correlations between the Neptune discovery chart and the birthchart of the United States. The position of Neptune at the time of his discovery was less than one degree from the U.S. moon. Neptune in the *U.S. chart* is at the "controversy" Mercury-Mars midpoint of the *Neptune discovery chart*.

Persephone in the *Neptune chart* is conjunct the Uranus of the *U.S. chart*. The Pluto positions in both charts are square to each other.

The sensational part of Neptune's discovery was that it had been found within one degree of Leverrier's prediction and less than 1.5 degrees from that of Adams. Regarding the intense controversy over who should get the credit for the discovery of Neptune, Adams or Leverrier, it was at first assumed that Neptune was indeed the planet they both predicted. But as observations were made, a new debate erupted. Neptune was not behaving as it should. The predictions of Leverrier and Adams, as well as Bode's Law, determined a mean distance of 36-to-38 a.u., but Neptune was found to be at 30 a.u. The predicted figures for eccentricity, period, and mass did not match the real planet at all. The astronomer Benjamin Peirce cause a sensation when he analyzed the figures, realized that Neptune could not account for the observed perturbations of Uranus, and pronounced that the discovery of Neptune "must be regarded as a happy accident."[83]

As we shall see in a later chapter, an almost identical situation was repeated in the story of Pluto's discovery. Not only were there the same disputed claims over who would get credit for the discovery, but also a similar controversy over whether the planet discovered was the same as the one predicted. If we step back from the noise of these professional debates, we can also see the repetition of the same implied message: the planets are not who we think they are, and there seems to be an extra one that's invisible and very disturbing.

BODE'S LAW AND THE DISCOVERY OF CERES

As early as 1596 the great astrologer-astronomer Johannes Kepler declared in his *Mysterium Cosmographicum* that an undiscovered planet must exist between Mars and Jupiter. In 1741 the German astronomer von Wolff suggested that something was not quite right with the spacing of the planets, and that perhaps one was missing. A few years later philosopher Immanuel Kant speculated about the problem. Finally, in 1772, a professor of mathematics Johan Daniel Titius published an arithmetic formula for such speculations in a footnote. The same year Bode quoted this formula almost verbatim in his introductory book on astronomy, postulating that another planet existed in the missing space.[84] The progression came out as follows:

Planet	Bode's Law in a.u.	Actual Distance in a.u.
Mercury	0.4	0.39
Venus	0.7	0.72
Earth	1.0	1.0
Mars	1.6	1.52
??	2.8	??
Jupiter	5.2	5.2
Saturn	10.0	9.5

If this progression is carried one step further, it yields a distance of 19.6 a.u. for the next planet. Uranus, at 19+ a.u., was close enough for Bode and others to hail Herschel's discovery of Uranus as final proof of Bode's Law. For ten years before this discovery, Bode had been promoting the idea that an undiscovered planet existed in the gap at 2.8 a.u. The fact that Uranus fit the scheme so well offered encouragement to astronomers to search for a new body. Finally, in 1801, the Sicilian astronomer Piazzi, who was making observations in preparation for a new star catalog, accidentally came upon what he first thought might be a comet. He followed it for a few weeks, then lost sight of it, but reported his discovery to Bode. Postal service being what it was in those days, it took two months for his letter to reach Bode, and then it was too late. His "comet" had moved too close to the Sun to be observed. Bode, however, realized this might be his long-postulated planet, and began spreading the word. Meanwhile, the philosopher Hegel had published an argument against the possibility of more than seven planets, his opinion being that it was foolish to predict another planet's existence merely on the basis of an arithmetic series.[85]

No one could find the lost planet until the mathematician Gauss set out to devise a reconstruction of the hypothetical planet's orbit based on Piazzi's observations. He succeeded in producing a tentative ephemeris which assisted the astronomer Olbers to re-discover the object exactly one year later, on January 1, 1802. This series of events of deliberate prediction leading to sighting is considered one of the landmarks in the history of planetary discovery, incidentally setting a precedent for future planet-hunting. (If we switch our minds to mythological "mode," we can imagine that the temporary disappearance of Ceres was another instance of her wandering in mourning, searching for her lost daughter; her re-discovery a foreshadowing of the reunion with Persephone.)

Piazzi named the planet Ceres, the tutelary goddess of his homeland of Sicily, and she was given her current astrological symbol by another astronomer who had been searching like Piazzi, but just missed successful sighting by a few days of bad weather. This astronomer, Von Zach, explained that "the symbol of Saturn represents a scythe, so the symbol of Ceres may represent a sickle, as Ceres is the goddess of corn and tillage."[86]   The distance of Ceres from the Sun, calculated at 2.767 a.u., fit neatly into the gap at 2.8 in the Titius-Bode progression.

When Ceres' discovery had been confirmed, astronomers believed that they had found the planet which corresponded to the gap in the Titius-Bode sequence, and were satisfied that the formula was correct. (This is not unlike what first happened with the discoveries of Pluto and Neptune.)

Ceres had been originally sighted by Piazzi at 23 degrees Taurus, but where Olbers re-discovered her was in the constellation Virgo. Less than three months later, Olbers was again looking for Ceres one night, and came upon a second object which he named Pallas, at 29 degrees Virgo (at the time Ceres was nearby at 24 Virgo). The discovery of this second planet, located at 2.67 a.u. (nearly the same distance from the Sun as Ceres), was an event that literally blew the minds of astronomers and made them re-think their hypotheses.

Olbers, an amateur astronomer, proposed a unique explanation. He believed that *asteroids* (the name given by Herschel), or *planetoids* (the name invented by Piazzi), or *minor planets* (as they were variously called), must be fragments of a blown-apart planet. His reasoning followed that all such broken pieces should pass through the same two (opposite) points in the sky, returning to the place of initial explosion-fragmentation. Olbers believed that these points of intersection in orbit would be the constellations Virgo and Cetus (just south of Pisces). Sure enough, Juno was discovered on September 2, 1804 in Cetus (3 degrees Aries), and Olbers himself discovered the fourth asteroid Vesta on March 29, 1807 in the constellation Virgo (astrological position 29 degrees Virgo).

A strong argument for Virgo as the sign of rulership of the asteroids is shown in this prominence of the sign and constellation Virgo in their discoveries, as well as in the fragmentation theory for their formation. 28/29 degrees Virgo was the discovery point for both Pallas and Vesta, and Virgo was also the location of the *re-discovery point* for Ceres (where Olbers found her one year after Piazzi's first sighting). Additional Virgo data: The perihelion longitude of Ceres at discovery was near zero degrees Virgo, and the North Nodes of Pallas and Juno are 23 and 20 Virgo, respectively.[87]

Of all the four major asteroids, it was Pallas that got the astronomers' minds to thinking. When the mathematician Gauss found that the orbits of Pallas and Jupiter were in a special ratio (18:7), called an *orbital resonance,* it was the starting point for many of his studies on orbit perturbations. The concept of inventive thinking associated with astrological Pallas is even extended to the symbology of the glyph of Pallas, which resembles the filament of a lightbulb. She does indeed represent the "lightbulb-over-the-head syndrome," the inspiration for ideas. Her capacity is strong in whole-pattern perception, which is precisely how Olbers responded to his discovery of Pallas; she inspired his theory that encompassed the overall origin of all the individual asteroids.[88] Although Olbers' provocative hypothesis has been generally discarded by modern scientists, the fact remains that it, along with Bode's Law, was instrumental to further discovery. As one historian of science admitted, "The strong point of [Olber's exploding planet theory] lay not in what it explained, but in what it had predicted."[89] This should serve as a lesson to scientists that discovery does not always require strict adherence to their properly validated, sanctioned theories.

Bode's Law was responsible for at least three valid predictions. Uranus and Ceres both conformed to its formula, and recognition of a similar law applied to satellites led to the discovery of Saturn's moon

Hyperion in 1848. While astronomers complain that Bode's Law is "not of precise physical significance," they have still been goaded by its challenge, to question whether the distribution of planetary (and satellite) orbits is due to chance, or to some "real" physical principle.[90]

It was during this initial period of the realization that many asteroids could exist (instigated by the discovery of Pallas), that major intellectual advances were made in the field of celestial mechanics that would affect further planetary discovery.[91] Laplace laid out his basic perturbation theory in great detail in 1799, and in 1809 Gauss published his innovative methods for computing orbits from a limited number of observations along with the method of least squares. Mathematical astronomers henceforth had sophisticated analytical tools for the problem of computing the motions of planets both known and unknown.[92]

PLANETARY CYCLES AND THE INTRUSION OF ASTRONOMY
ON THE FIELD OF ASTROLOGY

There are a multiplicity of stories about attempts to find planets, but there is one that is usually left out of the official renditions given by historians of Science, except to demonstrate the "frivolous" nature of astrology. In March of 1920, a man named P.W. Gifford, describing himself as a *Co-ologist* and a *Humanitarian Harmonist,* proclaimed the existence of two trans-Neptunian planets, which he had ascertained by means of Astrology. He identified them as Styxia, with a period of 228 years, and Janus, with a period of 426 years, adding that they were "presently 23 and 27 degrees in the astrological sign of Taurus."[93] Although Gifford's figures were totally in conformity with predictions being made at the time by astronomers, his attempt was quickly dismissed. As we shall see in the following chapter, Gifford's 228-year-period was closer to Pluto's actual cycle than any other prediction made (including that of the astronomer given credit for Pluto's discovery), a fact that has been totally ignored by the official historians.

The man who discovered Uranus, William Herschel, was an amateur in the eyes of his professional contemporaries; Clyde Tombaugh was a farmer's son with "only" a high school education when he discovered Pluto. Yet the Establishment of Science denies the role of amateurs and non-scientists in the process of discovery, a view that is supported by the successful record of the applications of Newtonian-based theory, and the belief that "celestial mechanics is entitled to be regarded as the most perfect science."[94] Science will not acknowledge the contributions of outsiders such as Velikovsky, or will downplay amateurs like Olbers, or will denounce astrology, while simultaneously usurping concepts and claiming such ideas as their own by restating them in revised language. This is currently happening in the so-called field of *Fractals,* which uses a repetitive pattern theory to explain Nature.[95] There are also the economists who have "discovered" a *Science of Cycles,*[96] and archaeo-astronomers who are describing ancient astrological texts as a product of *Calendaring,* the making of calendars to mark time.

These are but a few examples of the tendency of Science to make wholesale appropriation of astrological principles and disguise it with new labels and packaging. It is plagiarism, plain and simple, but astrologers would be hard put to prove such an accusation.

The fact is that there are many areas where astronomy overlaps with (or intrudes upon) astrology, and one of them is very pertinent to planetary discovery. What astronomers call *orbital resonance* has been long known to astrologers and mathe-magicians. It is the inter-relationships of the cycles of planets in our solar system, the rationale behind the pattern of repetitive conjunctions, such as the 20-year Jupiter-Saturn conjunction or the Neptune-Pluto 500-year-cycle. The orbits of Venus and the Earth offer a good example of how this works. Venus has a sidereal period of about 225 days, the Earth's is 365 days. The two planets have an orbital ratio of 3:5 (365 days multiplied by 3 is approximately equal to 225 days multiplied by 5). Venus completes five synodic periods of 584 days each in eight of our Earth years, and we get cycles of 8 years for Venus-Sun conjunctions. A table of such resonances is listed below.[97]

Saturn - Jupiter (28 yrs/12 yrs)	2:5
Uranus - Saturn (84 yrs/28 yrs)	1:3
Neptune - Uranus (165 yrs/84 yrs)	1:2
Uranus -Jupiter (84 yrs/12 yrs)	1:7
Neptune - Saturn (165 yrs/28 yrs)	1:6
Pluto - Neptune (248 yrs/165 yrs)	2:3
Pluto - Uranus (248 yrs/84 yrs)	1:3
Earth - Venus (365 dys/225 dys)	3:5
Venus - Mercury (225 dys/88 dys)	2:5
Jupiter - Mars (12 yrs/2 yrs)	1:6
Mars - Earth (2 yrs/1 yr)	1:2
Mars - Venus (2 yrs/225 dys)	1:3
Mercury - Earth (88 dys/365 dys)	1:4

Similar correlations of planetary orbital resonance connected with Sunspot cycles have been determined; examination has shown that "there seems to be a definite relationship between the orbital periods of the planets and the periodic nature of the sunspot cycle."[98] A capture technique model recently developed by two astronomers helps explain why planets seem to end up in certain orbital patterns, and this has been extended by another astronomer to explain how the outer planets accreted at locations where their orbital periods form simple ratios with those of their neighbors.[99]

COLLISION-ENCOUNTER ORIGINS OF THE ASTEROIDS, PLUTO AND PERSEPHONE

It turns out that orbital resonances (interactive planetary cycles) provide clues to the major unsolved problem of celestial mechanics: whether or not the solar system is orbitally stable.[100] Previous collisions within the system offer an explanation for these patterns of periodic conjunction.[101] In simple terms, it is a case of planets "returning to the scene of the crime." In fact, astronomers have been "discovering" that the solar system is not stable, but experienced previous internal collisions,[102] and of course they are careful to use their term *chaotic* and not *catastrophic* (the latter is too closely associated with Velikovsky and others who were really among the first to suggest such a concept). The chaos-catastrophe theory explains Pluto's possible former status as a moon of Neptune drawn away by a collision or near-approach of some Trans-plutonian planet. Two astronomers devoted to searching for Transpluto, Thomas C. Van Flandern and Robert Harrington, have postulated that irregularities in Neptune's satellite system, as well as the highly inclined orbit of Pluto, may have been caused by a single encounter of Neptune sometime in the past with a trans-Neptunian planet of considerable mass which was at that time in a circular orbit.[103]

Further speculations about the origin of the solar system and possible collision-produced alterations have been made by Zecharia Sitchin, a cult archaeologist who adapted Velikovsky's technique of interpreting ancient texts as the literal record of astronomical catastrophe. Because Sitchin is of the "Ancient Astronaut" school of thought, his ideas are classified by the Scientific Establishment as *crank,* but astrologers interested in Persephone and the asteroids should take a look at his book *The Twelfth Planet* and decide for themselves.[104]

Our own speculations arise from the apparent and multiple correlations of astronomy, astrology and mythology. It seems more than coincidental that Ceres, Venus, and Pluto, the three main characters in Persephone's myth, are not only astrological planets that share signatures in their respective discovery charts, but they are also the prominent planets involved in astronomical collision theories of origin of the solar system, which have been used to point to the existence of Persephone-Transpluto.

Chapter 9

SEARCHING FOR TRANS-NEPTUNE
REVEALS EVIDENCE OF TRANSPLUTO

"There is no reason to believe that ... this planet [Neptune] is the last one in the solar system."
— *Leverrier* (September 1846)

"Young man, I am afraid you are wasting your time. If there were any more planets to be found, they would have been found long before this."
— *Visiting astronomer to Clyde Tombaugh* (June 1929)

"I spent many years of my life blinking plates, and I know how easy it is to miss something."
— *Charles Kowall, discoverer of Chiron*

"The finding of Pluto was an important discovery, but what you did not find out there is even more important."
— *Astronomer Gerard Kuiper to Clyde Tombaugh* (1950)

Earlier in this book we mentioned that, when speaking of a new outer planet, the term "trans-Neptune" is technically a more accurate descriptor than "Transpluto." This will become clear in the following historical survey of planet hunting between the time of Neptune's discovery and that of Pluto. The survey will also show that astronomers have repeatedly found hard evidence proving there is more than one trans-Neptunian planet, a fact which has great relevance for our study of the astronomical reality of Persephone.

The first recorded mention of a planet beyond Neptune was actually made 12 years before Neptune itself was discovered. The astronomer Peter Andreas Hansen was quoted in 1834 as giving the opinion that there were two planets beyond Uranus.[105] The first publication predicting the existence of a Trans-Neptunian planet appeared only two years after Neptune's discovery. In this paper Jacques Babinet used simple arithmetic to derive the residuals of Neptune, and to conclude that another planet existed. He named it Hyperion, and proposed its distance as 47-48 a.u. and a period of 336 years.[106] Leverrier ridiculed this result, and no one apparently gave it much attention.

Leverrier, whose predictions led to the discovery of Neptune, also believed that another planet would be found, but worried that *"imagination played too large a part"* in the hypotheses of many astronomers searching for the planet.[107] There was, indeed, a great deal of imaginative thinking and subjective views expressed about the topic throughout the nineteenth century. For example, the mathematician Benjamin Peirce had applied his "law of vegetable growth" to the periods of the planets, believing it to be more accurate than Bode's Law.[108]

Although such theories are now scoffed at by scientists, labelled as quaint attempts by the misguided, and relegated to the status of minor footnotes in the history of science, they were once taken seriously enough to be published in professional journals (where the serious researcher probing dusty volumes can still find them today). If astrologers would like to watch astronomers squirm, they should simply ask them about astronomy's 50-year affair with the imaginary Vulcan, whose search was initiated by none other than Leverrier,[109] a fact that makes one wonder about his "too much imagination" statement. Another possibility is bringing up the subject of respectable Percival Lowell's predictions of intelligent life on Mars.[110]

But back to our story: Among those astronomers who followed the narrower path of traditional scientific method, many shared the view of Leverrier that perturbations of Neptune could be used to locate another planet beyond it. Another common line of inquiry was the evident relationship between comets and outer planets. (Since the aphelion distances of many comets are clustered around the orbital radii of outer planets, it was believed that such planets were diverting comets into elliptical orbits.)

One after another, astronomers came up with calculations to determine a new planet's position, mass, period, and distance from the sun. In

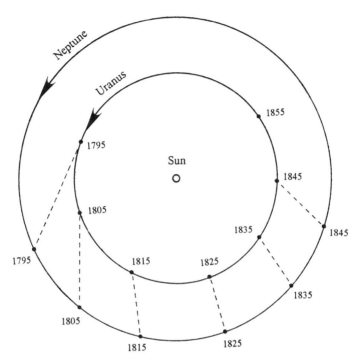

*Relative positions of Uranus and Neptune previous to the
discovery of Neptune in 1846.*

1877 David Peck Todd, then chief assistant of the U.S. Nautical
Almanac Office, figured the following specifications for the
hypothetical planet: 52 a.u., a longitude of 170 degrees, and a 375-year
period.[111] In 1879 Camille Flammarion suggested that an
undiscovered planet lay at a distance of about 48 a.u.[112] Georges
Forbes from Edinburgh, who was preoccupied (some said "obsessed")
with searching for this new planet for thirty years, insisted in 1880 that
there were two planets beyond Neptune (one at about 100 a.u., with a
period of about 1000 years).[113]

General scientific reaction to these proposals varied from wholehearted
acceptance to skepticism. One reason given by historians of astronomy
for this skepticism (besides the fact that searches conducted on the
basis of these proposals were unsuccessful) was the trans-Neptunian

planet's "strong attraction for cranks and visionaries."[114] Because astrologers are often lumped into that particular category, we might ask scientists what exactly do they use as criteria to distinguish "merely speculative and imaginative" from what is worthwhile research. The historic fact is that the trans-Neptunian hypothesis went through several cycles of popularity and respectability within the scientific establishment. (This, by the way, is still true today. Over the past few decades, the hypothesis of a Transpluto-Planet X has gone in and out of fashion.)

From the 1880s and into the twentieth century, there continued to be much activity in publishing about trans-Neptune,[115] with the theoretical bases for planet searches alternating between perturbation of known planets and the convergences of cometary aphelia. In 1900 Hans Emil Lau reviewed perturbations of Uranus and concluded that two planets had to be responsible, one at 46.6 a.u., with longitude 275 (+/- 180) degrees, and another at 70.7 a.u., with a longitude of 344(+/- 180) degrees.[116]

It is worth explaining at this juncture why predictions for hypothetical planets can be "plus-or-minus 180 degrees." Mathematical derivations of the position of an unknown planet from its gravitational effect on another planet yield two simultaneous solutions of 180 degrees apart (and therefore opposite each other). The reason for this is similar to why high tides occur twice a day, when the Moon is directly above our location (at the MC) and on the opposite side of the Earth (at the IC). We cannot tell by looking at the high tide whether the Moon is above or below us (180 degrees apart). Similarly, we can observe that something is "causing a tidal pull" upon a known planet, but we cannot tell at which end of the "axis-of-pulling" it lies. In the example of Lau's predicted position listed above, 275 degrees minus 180 equals 95 degrees (5 Cancer), which is incidentally about 6 degrees away from the position given in Hawkins' ephemeris for Transpluto. As readers will notice in this survey, many of the positions predicted for a trans-Neptunian planet correspond closely with the ephemeris positions for both Pluto and Transpluto.

Continuing with our story ... Gabriel Dallet wrote in 1901 that "Planet X" moved in an orbit of 47 a.u., with a probable longitude of 358 degrees.[117] Theodore Grigull, using the orbits of comets, calculated a distance of 50.61 a.u. and a longitude of 352-357 degrees.[118] Vicomte du Ligondes, using an entirely new method, found a predicted distance of 53.29 a.u.[119] Two other astronomers, Alexander Garnowsky and L. P. Choren M. Sinan, made further contributions to the literature on trans-Neptune.[120] The next major work on the problem came from the United States.

In 1908 William Henry Pickering announced he had found evidence of the existence of an undiscovered planet beyond Neptune. He was eventually to publish more on the subject than any other single investigator.[121] Pickering was born in Boston on February 15, 1858. His accomplishments in the field of astronomy were stunning. He led international expeditions to study solar eclipses, he erected the Flagstaff, Arizona observatory for Percival Lowell, he discovered Phoebe (the ninth moon of Saturn, which was the first satellite to be found by photography, and an anomaly because of its nondirect revolution around Saturn), and published the first complete photographic lunar atlas. As the photographic search for trans-Neptune continued without success, Pickering attributed the failure to a possible high inclination (more than 10 degrees off the ecliptic). His first calculations in 1908 determined a planet at 51.9 a.u. and a period of 373.5 years; by 1919 he had moved his "Planet O" out to 55.1 a.u., with a period of 409 years, and a longitude of 102.6 degrees. In 1928 Pickering reaffirmed an hypothesis he had first proposed in 1911, that the solar system should be increased by three undiscovered planets, what he called Planet O, Planet P, and Planet S. He incidentally considered that one of his planets (Q), with a projected mass equal to twenty thousand earths, was "practically a dark companion" to the Sun, orbiting at 875 a.u. with a period equal to the precessional cycle—26,000 years![122]

Meanwhile, astronomer Thomas Jefferson Jackson See had developed a new theory about planetary formation, which explained that planets closer to the sun would have more circular orbits than those further away. "In all probability," he wrote in 1909, "there are several more planets beyond the present boundary of the system, some of which may yet be discovered." He gave the distance of three such planets as 42.25, 56 and 72 a.u.[123] J.B.A. Gaillot reviewed the problem and admitted the possibility of two planets existing beyond Neptune, at distances of 44 and 66 a.u. and a longitude of 271 degrees.[124] Hans E. Lau also studied the matter from a different perspective dependent upon mass, and found that the only place the hypothetical planet(s) could be located to be in longitudinal sectors 30-to-120 degrees and 210-to-300. He also noted that there "may exist two or more major unknown planets beyond the known limits of the solar system."[125] Lau later adjusted his figures to be two trans-Neptunian bodies at 46.5 and 71.8 a.u.[126]

The next big step in researching trans-Neptune was taken by Percival Lowell, whose efforts would eventually lead to the discovery of Pluto. Lowell was born in Boston on March 13, 1855. Although he began working on calculations for a new planet in 1906, his first publication on the subject did not appear until 1915.[127] Like many previous investigators, Lowell found two possibilities of position for the planet, 180 degrees apart, at 84 and 263 degrees, with distance between 43 and 44.7 a.u. He also believed the planet would have a high inclination of about 10 degrees. Using photography, he searched for the planet for ten years, until his death in 1916. He even specified in his will that the search should continue at Flagstaff.[128]

Meanwhile, new predictions were being made. In 1921 Theodore Grigull proposed new positions for two trans-Neptunian planets, one at 46.784 a.u. and another at 90 a.u.[129] Emile Belot presented a theory that placed a new planet at 62.44 a.u., later revising it to 56 a.u., with a period of 407 years.[130] Finally, after a 12-year lapse following the death of Lowell, the search for trans-Neptune was renewed at the Lowell Observatory by Clyde Tombaugh, under the direction of Vesto Slipher. Using a blink microscope, Tombaugh examined photographic plates and discovered what we now know as Pluto. Slipher had waited to announce the discovery, choosing March 13 because it was the 75th anniversary of Percival Lowell's birthdate and the anniversary of the discovery of Uranus. The symbol PL was adopted to further honor Percival Lowell with his initials.

Once Pluto's orbit was determined, several observatories found they had taken pre-discovery photographs (Mt. Wilson in 1919, Konigstuhl

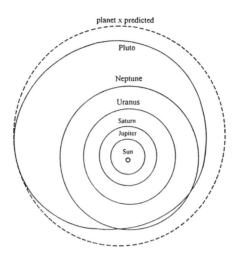

Lowell's predicted orbit of Planet X compared with the observed orbit of Pluto.

in 1914, Yerkes in 1921 and 1927, and Uccle in 1927).[131] Such photos helped correct orbit determinations for Pluto. Many astronomers believed at first that Pluto was the trans-Neptunian planet everyone had been looking for, but further analysis proved otherwise. The month after Pluto's discovery was announced, the astronomer E. W. Brown rejected Lowell's and Pickering's predictions as worthless to the actual discovery, claiming that it was a complete coincidence.[132] This view has since been confirmed repeatedly and is widely accepted by most astronomers.[133]

Some of Pickering's predictions for Planet O were much closer than Lowell's Planet X, but it was Lowell who received the posthumous fame. Pickering was the only one of all the predictors to suggest that one planet could penetrate the orbit of another (as Pluto does with Neptune). Even after the official discovery of Pluto, Pickering continued to publish articles debating Lowell's figures well into 1931.
When Pluto was discovered, he was at 39.29 a.u., 246 year period, and 108.5 degrees longitude. For the record, the closest that Lowell's and Pickering's predictions came to the actual orbital elements of Pluto

were: Lowell's Planet X (1915) was 43 a.u., 282-year period, and longitude (corrected to 1930) 102.7 degrees, and Pickering's Planet O (1919) was 55.1 a.u., 409 years, and 102.6 degrees longitude. Comparing predictions with the real figures, there were gaps of almost 6 degrees off position, 3-to-15 a.u. off the distance, and 36-to-163 years off the cycle. These differences give a strong suggestion that Pluto is not the only trans-Neptunian planet. It is certainly clear that Pluto's position does not conform to where the astronomers' formulas said a trans-Neptunian planet should be.

The reader may have noticed in this historical survey that every one of the astronomers' predictions made for a trans-Neptunian planet placed it much further out than Pluto's actual orbit. (Yet scientists have the gall to criticize astrology for having a poor record for "correct" predictions!) This fact, that all the predictions located Trans-Neptune at a greater distance than where Pluto was discovered, is one of the most potent scientific indicators of the existence of Transpluto.
In their efforts to search for planets, astronomers might well heed the advice offered by Clyde Tombaugh, who devised the following:

The Ten Special Commandments for a Would-Be Planet Hunter
by Clyde W. Tombaugh, the Man who Discovered Pluto[134]

Behold the heavens and the great vastness thereof, for a planet could be anywhere therein.
Thou shalt dedicate thy whole being to the search project with infinite patience and perseverance.
Thou shalt set no other work before thee for the search shall keep thee busy enough.
Thou shalt take the plates at opposition time lest thou be deceived by asteroids near their stationary positions.
Thou shalt duplicate the plates of a pair at the same hour angle lest refraction distortions overtake thee.
Thou shalt give adequate overlap of adjacent plate regions lest the planet play hide and seek with thee.
Thou must not become ill in the dark of the moon lest thou fall behind the opposition point.
Thou shalt have no dates except at full moon when long exposure plates cannot be taken at the telescope.
Many false planets shall appear before thee (thousands of them), and thou shalt check every one with a third plate.
Thou shalt not engage in any dissipation, that thy years may be many, for thou shalt need them to finish the job!

Chapter 10

AFTER PLUTO'S DISCOVERY:
THE HUNT CONTINUES

The discovery of Pluto did not stop the astronomical search for Trans-Neptune; in fact, it urged and encouraged astronomers to continue the hunt for an unknown planet. In this chapter we shall survey the history of planetary discovery since 1930. We will be paying particular attention to predictions that coincide with the tentative Transpluto ephemerides used currently by astrologers.

Even with the discovery of Pluto, many astronomers didn't skip a beat in their continued search. Pickering and Tombaugh, who figured so highly in the discovery, kept right on going. Among the numerous planets proposed by Pickering, the one that should interest those searching for Persephone is his Planet P, for which he gave (the 1931 postulation) 75.5 a.u. and a period of 656 years. Pickering's Planet P was then positioned at 28 degrees Capricorn, exactly opposite the *Hawkins' ephemeris* position of Persephone at 27.9 degrees Cancer. Pickering did not believe that all his planets were discoverable, but among the ones he considered possible candidates for discovery was P.[135]

Less than three months after he had discovered Pluto, Tombaugh was instructed by Slipher to continue searching the sky for other planets. During the 14 years that Clyde Tombaugh surveyed the sky in search of trans-Neptunian objects (1929 to 1943), the astrological planet Neptune was situated in tropical Virgo. Speaking of nit-picking and detailed work, Tombaugh examined during this time over 45 million stars. During his 14 years, he found several new star clusters, one new comet, about 775 new asteroids, and one new planet.[136] Most astronomers have so much respect for his work, that many feel he could not have missed sighting a transplutonian planet (assuming it was within the 70% of the sky Tombaugh searched and was bright enough for him to spot, at least 16.5 magnitude).[137]

There remained too many questions "out there" for astronomers to be satisfied that they could stop looking. Too many predictions had placed trans-Neptunian bodies at a distance further than Pluto. Lowell himself had suggested a second body at about 75 a.u., although he had thought it would be futile to search for it. The sequence of events in planet-hunting went like this:

In 1942 Robert Richardson proposed a tenth planet at 36 a.u. to explain the delay of Halley's Comet in reaching perihelion.

In 1946 M. Emile Sevin from France predicted a tenth planet at 78 a.u., based on the perturbations of Neptune.[138] He had calculated a period for this hypothetical body by an unusual method of dividing the known planets into groups of inner and outer bodies. He came up with a possible trans-Plutonian body with a period of 677 years. He subsequently derived a distance of 77.8 a.u. and a period of 685.8 years, but his prediction stirred little interest among astronomers. We shall see later on that Sevin's work did interest astrologers, however, and eventually led to the calculations of the *Hawkins ephemeris* for Transpluto.

In 1950 K. Schuette from Germany proposed a tenth planet at 77 a.u., based on comet orbits.[139]

In 1954 H.H. Kritzinger from Germany refined Schuette's work and predicted a tenth planet at 65 a.u. with a period of 524 years; in 1957 he revised his work, predicting a planet at 75 a.u. with a period of 650 years; and finally, in 1959, re-analyzed his work and came up with 77 a.u. (Schuette's original distance) and a period of 675.7 years, a planet not unlike Sevin's and somewhat similar to Pickering's Planet P.[140]

Throughout the post war period, astronomers were becoming more knowledgeable about the orbit of Pluto, and this meant better calculations for Transpluto's orbit. In the 1960s they were anticipating that Pluto would start becoming an important factor in the search for a transplutonian planet because he was going to be crossing inside the orbit of Neptune in the 1970s and reaching perihelion in the 1980s. They knew it was going to be possible to more accurately figure his mass. More exact knowledge about Pluto's mass and orbit would in turn determine a better understanding of his contribution in perturbing the orbits of the other planets, and this would ultimately result in a better picture of how much "perturbing" residual was left over to assign to the hypothetical planet and thus "define" where it was.[141]

Astrologers have given some thought to the meaning of Pluto being inside Neptune's orbit (from Jan. 21, 1979 to March 14, 1999),[142] but they could have a real picnic interpreting this fascinating analogy from the standpoint of the Persephone myth. We may surmise that the closer Pluto gets to Neptune, the more we can understand how he disturbs the astrological functioning of the outer planets, and out of this realization we come to understand Persephone. Even more interesting is the image of Pluto crossing into the more familiar (closer) world, as he does in the myth to abduct Persephone.

In 1970 Edward J. Gunn predicted the unknown planet to be currently at 10 degrees Leo. (The *Hawkins ephemeris* shows Transpluto at that time to be 13.5 degrees Leo.) This was basically in agreement with Sevin's figures, but by an independent method. Interestingly, Gunn used a conjunction of Neptune with the hypothetical planet in 1884 to figure out that Transpluto must move 0.92373 degree per year. He also noted that the multiple conjunctions of Jupiter, Saturn and Neptune during that period of the early 1880s made it difficult to isolate one source of perturbation.[143] Astrologers can certainly sympathize with this problem, namely how to isolate the influence of one planet from the wider impact of a stellium or multiple conjunction.

In 1972 Dennis Rawlins supported the importance of the 1795 pre-discovery observations of Neptune as a key to perturbations of its current orbit. He consequently worked out that "the only satisfactory solutions found were for bodies whose orbits lay between 50 and 100 a.u. from the Sun, and whose positions at Jan. 1973 lay within a relatively small sector of the ecliptic."[144] Depending on the mass of the hypothetical planet, it could be predicted at 75 a.u. and 315 degrees longitude, according to Rawlins. Recalling that longitude figures can be +/- 180 degrees, we note that this location could be 135 degrees, which corresponds with *Hawkins'* Transpluto *ephemeris* position in 1973 at 15 degrees Leo.

Joseph Brady shocked the astronomical community in 1972 by predicting a planet at 60 a.u. that moved in retrograde orbit with a mass 285 times that of the Earth and a period of 464 years. The reaction to this proposal within astronomy was very negative.[145]

Gleb Chebotarev from the Soviet Union predicted two undiscovered planets in 1975, at 54 and 100 a.u., based on an analysis of comet orbits.[146]

From 1977 to 1984 Charles Kowal conducted a systematic search for undiscovered bodies in the area 15 degrees on either side of the ecliptic; in the process he found five comets and fifteen asteroids, including Chiron, which initial press coverage treated as a "tenth planet." Kowell himself made an attempt to calculate a location for a new planet, and came up with numbers very similar to those of Harrington and Van Flandern. Between the extensive searching of Tombaugh and Kowal, it seemed to many astronomers that the hypothetical planet should have been sighted. But as Kowal admitted, "I spent many years of my life blinking photographic plates, and I know how easy it is to miss something."[147]

On June 22, 1978, Pluto's Moon was discovered and named Charon by James Christy. This was not an isolated discovery, for it made possible the determination of Pluto's mass and speculation about the origin of Pluto, both of which influenced the theories about Transpluto. Pluto's mass was now known to be far too low to fully account for the orbit perturbations of Uranus and Neptune (which had originally prompted the successful search for Pluto itself). The discovery of Charon and the determination of Pluto's very small mass made it certain that Pluto could not be Percival Lowell's predicted Planet X.[148] In fact, the existence of Transpluto began to seem like the best explanation for a lot of peculiar characteristics about Pluto and its moon.

In 1979 Robert Harrington and Thomas Van Flandern proposed a tenth planet in a highly inclined and eccentric orbit at a distance of 50-to-100 a.u., and with a period of about 800 years, to explain the irregularities in the motion of Uranus and Neptune, the unusual orbits of Neptune's moons, and the anomalous orbit of Pluto. Their idea was, that if Neptune originally had a well-ordered system of three satellites, but experienced an encounter with a single unknown planet (with a mass 5 times the earth), the disruption could have knocked Pluto loose and altered the orbits of the remaining Neptune moons, also ripping off a small chunk of Pluto to become his moon. This "marauding planet" would have been thrown into a more elongated orbit, ending up with a distance and period equivalent to the figures they proposed. Harrington and Van Flandern guessed the location to be in the constellation Capricorn or Aquarius (from about 310 to 345 degrees longitude). Subtracting 180 degrees gives a possible location in Zodiacal Leo, as in the *Hawkins ephemeris*. Harrington commenced a search for the planet based on this theory.[149]

During the 1980s, astronomers explaining transplutonian planets began to lean more and more in the direction of close-encounter theories, like that of Harrington-Von Flandern as well as the idea of a solar companion Nemesis.[150] Two researchers, Donald Morris and Thomas O'Neill, considered this theory, concluding that such a solar companion might exist only if its closest point of orbit were beyond Neptune. Such limits on close encounters reflected back on theories about the possibility of a major planet beyond Neptune.[151]

In 1985 Daniel Whitmire and John Matese proposed a tenth planet to explain periodic mass extinctions on earth due to comet showers, with an orbit of about 80 a.u., a substantial ellipticity and inclination, and a period of 700 years. Matese and Whitmire noticed that their hypothetical planet was quite similar to that of Harrington and Van Flandern. The original idea of extinction of dinosaurs had been published by Luis Alvarez and others in 1980; it proposed that the impact of a comet had thrown so much dust into the earth's atmosphere that the sunlight was blocked, and plants and animals died.[152]

In 1987 John D. Anderson analyzed the Pioneer 10 and 11 spacecraft trajectories for evidence of perturbations. He found none, but concluded this was because the suspect planet was highly inclined (30 degrees tilt from the ecliptic) and probably elliptical in orbit, so it could be too far away to affect the spacecraft, but still would account for the discrepancies in the motion of Uranus and Neptune. Anderson proposed a planet with a period of 700 to 1000 years.[153]

Conley Powell predicted in 1987 that the position of a tenth planet would be in a nearly circular, low inclination orbit based on the old-fashioned method of analyzing Uranus residuals. Like many planet-hunters, Powell discarded the use of Neptune residuals from consideration because Neptune had not yet been observed for one complete revolution. In astrological interpretation, we might say that we don't yet understand Neptune (the Piscean principle of sacrifice-suffering) well enough to be able to comprehend Persephone. Powell predicted an object at 60.8 a.u., with a period of 494 years. He placed it at first in the constellation Gemini, but later re-examined his data and fixed Transpluto much nearer to Pluto, with a nearly similar distance and period. If his planet is found, he says he will name it Persephone.[154]

In 1988 Thomas Chester and Michael Melnyk completed a two-year review of images from the IRAS (Infrared Astronomical Satellite) covering 1/10th of the sky in search of the tenth planet; their original plan was to study one-half the sky.

In 1989 Rodney Gomes published his calculations for a new planet using perturbations of Uranus and Neptune, and came up with several plausible locations, including one with a nearly circular orbit at 45 a.u., and another with an eccentric orbit at about 80 a.u. Gomes believed that the most likely current position would be in the constellation Cancer or Gemini, nearly opposite that of Harrington's favored location, and corresponding to the *Hawkins ephemeris* position for Persephone in Leo.[155]

There are some astronomers who feel there are too many problems with the data on Uranus and Neptune to use them for calculations of a new planet. Among them is E. Myles Standish of Jet Propulsion Labs, who believes there is no need for a tenth planet to explain residuals because he thinks there are too many possible errors in earlier observations. In other words, it is not essential that a tenth planet exist in order to account for the motions of the other planets. Even Standish admits, however, that there is no way to rule out absolutely the existence of a tenth planet.

What do all these astronomical theories indicate for our understanding of Persephone? One astronomer has given us some clues. We quote from him directly because what he says seems to neatly describe the meaning of astrological Persephone, albeit in astronomers' lingo.[156]

No object in the solar system has caused more trouble. Imagine a Most Wanted poster for this fugitive planet:

WANTED on cosmic charges:
 * Disturbing the motion of Uranus and Neptune.
 * Smuggling short-period comets into the inner solar system.
 * Suspected of trespassing at Neptune, driving Triton and Nereid berserk, and kidnapping Pluto.
 * Repeated assaults on Earth with deadly comets, causing periodic mass extinctions of life.

DESCRIPTION of fugitive: One to five Earth masses; eccentric, with odd inclination; likes to leave subtle clues to tantalize astronomers; lives in trans-Plutonia, constantly on the move, no known address, might repeat movements every 700 years; knows how to hide.
NOTE: Substantial reward for information leading to his arrest.

THE LATEST SCIENTIFIC RESEARCH ON TRANSPLUTO

The early 1990s have witnessed a dramatic change in the way astronomers have viewed the problem of what is really orbiting beyond Pluto. In 1993 astronomers discovered a few "objects" at about 50 a.u., a territory which may (according to the latest theories) be a source area for comets, asteroids-turned-into-comets, "rogue planets," and other solar system oddities. This most recent evidence seems to point to the possible existence of a large number of small, dispersed objects rather than one larger (planet-sized) solid body beyond Pluto.

This latest scientific evidence and theorizing does not necessarily conflict with the astro-mythological ideas put forth in this book. One reason is that hypotheses about a Transplutonian planet have gone consistently in and out of fashion for about 150 years, so we must always practice caution in embracing the latest theories. Another important point to remember is that there exists a relevant precedent, the asteroid belt, which astrologers often treat in principle as one unified "body," even though this is in obvious contradiction with the fact of their physical, fragmented multiplicity.

See Ron Cowen, "Plutos Galore: Ice Dwarfs May Dominate the Solar System's Planetary Population," Science News (Sept. 21, 1991), pp. 184 ff.; "Neptune's Friends," New Scientist (Oct. 2, 1993), p. 12; Freeman Dyson, "Hidden Worlds: Hunting for Distant Comets and Rogue Planets," Sky & Telescope (January, 1994), pp. 26-30.

Chapter 11

THE ASTROLOGICAL DISCOVERY OF TRANSPLUTO, AND WHY WE SHOULDN'T CALL HER BACCHUS

Astronomers are not the only researchers who have investigated the idea of a Transplutonian planet. Several astrologers have also discovered or invented hypothetical planets that correspond to the astronomers' predictions, assigning positions, cycles, and various names to such planets. This short chapter will present some, but not all, of these astrological predictions. Because astrology as a body of knowledge is not as organized, referenced, monitored, nor as widely published as is astronomy, it is relatively difficult to track down all the obscure astrological writings that contain such information.[157]

For the same reason, it is almost impossible to determine the exact number of hypothetical planets devised by astrologers, but the figure was approximated in 1976 as "between 30 and 50."[158] According to the same source, ephemerides have been published for at least 25 hypothetical planets. Some of the best known are the "transneptunians," originated by the Hamburg School in Germany, studied by the Uranian School astrologers in the United States, and named Cupido, Hades, Zeus, Kronos, Apollon, Admetos, Vulkanus, and Poseidon.[159] Other hypothetical planets, such as Lilith and Vulcan, have received some attention in astrological publications,[160] but we shall limit our discussion to those whose predicted cycle, position, and distance from the Sun most closely correspond to the astronomical calculations of a Transplutonian planet. In other words, we will look at some of the astrologically-predicted planets whose locations range from 40 to 100 a.u., with cycles of 3-to-10 centuries.

As the astrologer Charles Jayne pointed out in his publication *The Unknown Planets*,[161] many of the so-called hypothetical planets have periods and distances that correspond to predictions made by astronomers (such astronomical predictions were discussed in previous chapters). The table below is based on Jayne's summary and lists these planets.[162]

Planet	Period in yrs.	Distance in a.u.
(Pluto)	244	39.3
Cupido	270	42.1
Pan	336	48.7
Hades	355	50.6
Isis	377	52.7
Zeus	426	57.3
Chronos	484	62.4
Apollon	552	68.9
Admetus	591	71.5
Morya	634	75.1
Samson	680	78.7
Poseidon	732	82.8
Hermes	842	90.5
Osiris	986	101.3

What is interesting about this list from a mythological standpoint is that more than half the planets have names of figures involved in the Persephone myth, or variations thereof. (The Isis-Osiris myth is the Egyptian version of this story; coincidentally, the midpoint of their two cycles is 660 years. Cupido is a variation on Eros, the sidekick of Venus. Pan is a very Dionysian character. Zeus, Hades and Hermes are the Greek names for Jupiter, Pluto and Mercury, all of whom were featured prominently in Persephone's myth.) It would appear that astrologers, like astronomers, have recognized the appropriateness of choosing character names from Underworld myth for Transneptunian planets.

The periods for Admetus and Samson (Vulcanus) were altered to 617 and 662 years; Morya's period has been recalculated to 639 years.[163] Another astrologer, Maurice Wemyss, postulated a planet called

Hercules, which completed one revolution in approximately 654 years, and which would be located currently in Leo.[164] Larry Ely, the astrologer who assisted in producing the ephemeris contained in Hawkins' book *Transpluto,* has been working to refine his Persephone Ephemeris since the mid-1970s, and his figures, at last report, correspond to the 670-685 year range.[165]

As Jayne noticed, we have too many postulated planets converging on one range of cycle, and the figure is increased when we add on the list of astronomical predictions. We recall from previous chapters that the astronomer Pickering gave his Planet P a period of 656 years, and that Bode's Law indicates a planet at about 77 a.u. with a period of 678 years. The calculations by astronomers that led to the publication of Theodore Landscheidt's ephemeris for Transpluto,[166] giving a period of 685+ years, are outlined in Hawkins' *Transpluto,* as well as in Jayne's *Unknown Planets.* Landscheidt, Ely, and Hawkins place Transpluto currently in Leo, as has Edith Wangemann from Germany, though she re-dubbed the planet Isis.

It seems clear that astrologers, like astronomers, are in some general agreement over the whereabouts of a transplutonian planet, despite their individual differences over details. While we do not wish to dismiss the value of pinpointing and correcting the orbital elements, we do feel that more focus has been placed on finding "the right answer" to technical questions about Transpluto than to addressing the broader and more contextual issue of "what it all means."[167] As we explained in earlier chapters, the path to understanding Persephone is not solely through the familiar (and masculine) habits of the analytical, scientific mind-set. Few attempts have been made to interpret the planet Transpluto astrologically, and they are all based on a confusion over mythological identity.

DEMYSTIFYING ASTRO-MYTHOLOGY

John Hawkins introduced Transpluto as Bacchus, gave "him" the rulership of Taurus, and assigned the "winking Mars" symbol which is currently used by many astrological software programs. As we have attempted to show throughout this book, Dionysos (Bacchus) is a relevant factor in Persephone's myth, and it is easy to confuse "who's who" when correlating astrology with mythology. (Even the Greeks and Romans themselves tended to cross-associate and interweave the myths and characters of the Persephone myth. Bacchus was described sometimes in their literature as interchangeable with Ceres.)[168] The astrologer Andrew Bevan has continued Hawkins' Bacchus-and-Taurus interpretation for Transpluto in his recent publications.[169]

Such interpretations do suffice as adequate explanation for what is observed through surface examination, for the myths and associations of Dionysos do seem to qualify as determinants of wounding, suffering, sacrifice, altered consciousness, fertility, ecstasy and freedom, immortality and resurrection.[170] There are two serious questions we face, however, when we delve deeper: *Are these really the workings of a masculine god,* and *does the rulership of Taurus really make sense?*

Regarding the confusion over Transpluto's rulership of Taurus-Libra, there is a constant need to distinguish Venus and Persephone. As with all separation dilemmas, the important factor is *degree of consciousness.* Venus is less conscious than Persephone, and the Moon (exalted in Taurus) is even less conscious than Venus. Gaia may be the least conscious of all (she is pure primitive creativity). Much of Persephone's myth that is commonly given attention is full of all the Taurus-related goddess images of security and bondage to the material plane. Popular versions of the myth often stop with the stage of her return to her mother, as though this "final" return represented the achievement of greater consciousness in one static event. We do "return to the mother," whether it's Gaia, Luna, or Venus, the state of relative unconsciousness, only to be pierced again and drawn toward the underworld, "forced" to face Pluto and upon our return, develop our conscious state.

One of the most insulting aspects of the patriarchal period of the past 3 or 4 millenia (the Pisces and Aries Ages) has been the deliberate and limiting definition of the feminine as either the sneaky wiles and temptations of Venus or the smothering lunacy of big Mama Moon. But even if we give those inner feminine planets the most due they deserve, they are still limited to functioning on personal and social-moral levels. It takes a further-out perspective of Persephone to see all the intra-Saturnine personal planets in one glance. How tiny and contained they look from the transpersonal viewpoint. All the outer, discovered planets (Uranus, Neptune, Pluto) rule signs that are Zodiacally beyond the personal ego development phase represented by the sign sequence Aries through Virgo. Why would we assign this latest outer planet to Taurus, located in the Zodiacal area ruled by personal planets?

A RE-EVALUATION OF SUFFERING

There are several other points of criticism for the Bacchus interpretation. Dionysos was, like Venus-Aphrodite, a god-come-lately to the Greek Pantheon; so "late," in fact, that he is considered to be emblematic of the transition into patriarchal Christianity and its wounded, suffering, dying and resurrected Christ. What this means is that Dionysos and his myth are riddled with non-Pagan associations, white-washed and adulterated with the pseudo-morals of Christian tradition (those which we are still trying to sort out). As the astrologer Rob Hand has pointed out, in the early days of Christianity, when the Gnostics were rejected, a decision was made to chose a conservative, Old Testament view of the Godhead as a Saturnian entity with an Ego who said "I am a jealous god; thou shall have no others before me." The Gnostic view, which was and is still considered heretical, recognized that Christ by his suffering was saying "When I suffer, you suffer, and god suffers too. We are all god."

We have been told that Christ's last words on the cross were "my god, my god, why hast thou forsaken me?" (a statement that focuses on loss, abandonment, and ultimately results in a mass-consciousness archetype of the exaltation and worship of the innocent, abandoned one who suffers.) But the words actually translate as "my god, my god, for this I came into being."[171] By dying on the Cross of matter, Christ showed that *god had chosen voluntarily to experience suffering.* And this means that humanity can choose also, which is closer to the message of the original myth of Persephone, in which she goes willingly into the Underworld.

By focusing on the negative connotations of *sorrow,* whether it be Christ's or Dionysos' sacrifice, or the rape-abduction-piercing of

Persephone, we focus on loss and limited manifestation. We need to turn this around and apply the "mirroring" technique. Persephone is not the wounded one, but the one who wounds. She slays — the Greek word *phone* refers to murder or slaughter — it's in her name. She pierces the ego so that we connect with the heart. *Breaking the heart* (separation from a loved one) is one of the possible images. We merge with ourselves through the process of piercing or slaying: the old forms of consciousness are broken open. Life is indeed heart-breaking ("All life is sorrowful," says the Buddha), but this is how we grow.

In order to move through the final stages of the Piscean Age, we must open up our concept of suffering to include the renewal that follows it, as rebirth follows death. We may fear this process because we have been taught to believe that to accept suffering in general means that we will become callous and unfeeling, oblivious to the pain of others, and consequently inhumane; or that to accept suffering means that we would just give in, roll over and die. The only thing that really dies when we accept suffering is our ego! Psychotherapy theories and popular moral norms in current fashion are riddled with distorted ideas about suffering, namely that people are *wounded* and that this is something bad that's got to be fixed.

"Wounded" as such is simply the eternal human condition; it is not something that can be healed (fixed). An incredible amount of energy is currently being focused on seeing wounds as justification for behavior ("I married a child-beater because my Dad beat me."), and as exaltation of the innocent or the victim. People can wallow around in therapy for years, ostensibly to "heal" these "wounds." (These are the same people that confront their "happier" friends with the advice: "You're actually *in denial*. You haven't really *mourned* enough," — and with the wisdom of a dilettante, they quote the case of Ceres needing to mourn for Persephone's descent.) With this kind of attitude, we're stuck as a society in the moral dilemma over how to prioritize who is suffering the most (do we use taxes to house the homeless, rehabilitate criminals, support welfare mothers, or what?).

We see how this confusion manifests in writings that attempt to correlate mythology with astrology. People start projecting their own supposed wounds unto the archetypal images. In Jennifer Woolger's

117

The Goddess Within, Sylvia Plath is given as an example of the "Persephone archetype." This idea was further developed in a recently published astrology article, in which the author says that Plath was "caught in the end of the old cycle — death without the vision of a way through to rebirth — Persephone trapped in the underworld without having fully made her descent or subsequent journey home."[172] What this author is doing (in the name of astro-mythology correlation) is taking a myth and revising it by projecting our own problems upon it. Persephone is never trapped; she always returns. The myth clearly states that it is Pluto without Persephone who is death without rebirth.

We find similar confusion in another piece of astro-mythological writing, where the author states that the re-emergence of the feminine into our consciousness is "fraught with difficulties because the goddesses are re-emerging in their wounded form."[173] Herein lies the error of illusion: mistaking the shadow or the appearance of the form (humans calling themselves "wounded") for the content (of the myth itself). There is nothing inherently wounded or deficient about the goddesses themselves. The Goddesses don't change — societies and values do.

This is not a simple argument over interpretation. Free-association is a fine method when used appropriately to transcend old patterns of thinking. What we must be watchful of, in our enthusiasm to "embrace" mythology, is the tendency to corrupt or dilute the positive images and their powerful, ancient truths with projections from our own distorted reality. Mythology provides guidance models for us to look up to; the myths are "how-to-do-it manuals" for life-skills. If we alter myths to reflect our current reality (which is basically what modern pop culture and education is attempting), we shouldn't be surprised if people squawk about crime and moan about how there's no values left in society. We can formulate our lives around myths, not vice versa. As it has been said, *the old ways are the good ways.*

The fact is, we're all suffering because that is a given human condition ... it comes with having a body. But this idea is in conflict with what contemporary religion-morality has taught. Up until the modern era, the Western Judeo-Christian tradition taught that Heaven belonged to the afterlife. This was replaced during the Scientific and Industrial

Revolutions by Calvinism, which teaches that Heaven can be found (or rather, bought) on earth ... If you work hard enough and make enough money, it shows you are one of the chosen. It comes as a great shock to people to realize that, no matter how many social reforms are enacted, no matter how many therapy session they attend, and no matter how much material wealth they produce, *suffering still occurs.*

The key is in seeing the value of suffering, pain, wounds, or the piercing of Persephone as necessary and functional to growth and development. It is because the ecocycles have been disturbed and the earth raped that environmental consciousness has been awakened. It is because the Chinese waged an ethnic war on Tibet that the Dalai Lama and other monks were forced to migrate and consequently spread a spiritual message to the Western world. Yes, our immediate reaction is to be appalled. We must recall the message, "Quiet, Inanna, the ways of the Underworld are perfect." (Christians put it this way: "The Lord works in mysterious ways.") We can *see* in a new way and recognize the ultimate function of suffering, not simply our immediate emotional reaction. We perpetuate our own pain by hanging onto the illusion that suffering is wrong, that wounds must be healed, that life can be fixed, and that before any change can happen in our lives, we must process our "stuff" to the point of final resolve. There is no such perfected state on this planet. We can all still go out and help others or develop our selves without insisting on first perfecting or fixing our own wounds.

Approaching life as if each one of us is in some way wounded and is lacking, having scars or gaps that must be healed or filled, simply keeps us in a rut. When we realize that everyone has their own private hell, their own personal prison, and that we all *share* this as a human condition, we can dismiss the specialness we like to attach to our own personal brand of suffering, and expose the fraud for what it is: something which separates us from others and puts one person's suffering in competition with another's. When we abandon our own egotistic child, we can get on with treating everyone (including ourselves) as mature and whole. Just as Demeter-Ceres lets go of her attachment to her co-dependent relationship with an immature Persephone, releasing the old bondage patterns and allowing a "space" for Persephone to return as a grown woman, we all can cross the great border of the horizon, from the Ascendant to the Descendant. Then we enter into the Persephone-ruled sign of Libra, finding a passageway from the "Self" signs into all the "Other" signs of the Zodiac, and embracing their true nature, from the conscious relationship of Libra to the wholism and acceptance of suffering in Pisces.

Chapter 12

PERSEPHONE IN THE CHART
AND IN OUR LIVES

Those who practice astrology will want to know how to "use" Persephone in the interpretation of the birthchart and mundane events. There is an ephemeris at the end of this book to help plot birth positions and transits. Readers are also referred to the ephemeris published in John Hawkins' book *Transpluto,* and are encouraged to consult computer programs that include Transpluto-Persephone. As stated previously, such ephemerides are tentative, based on the "best guess" of astronomers' calculations, and will someday be corrected when the planet is actually sighted. For now, this is what we have to work with, and the positions indicated do seem to make sense astrologically.

We can look to the natal position of Persephone, her sign and house placement, aspects to other planets, transits, and progressions, just as we do with the known planets, and apply what has been learned from analysis of her myth. In looking to the house Persephone occupies natally, we would see powerful transformative relationships arising in this area of life. We find her very active as an indicator of change in relationships.

Before we proceed to examples, an explanation is in order. Readers may have noticed that this book's treatment of Persephone does not include detailed discussions of "famous people's charts" so common to most astrology texts. (Other astrologers have used the charts of Sylvia Plath and Patty Hearst to illustrate the Persephone-type.) It is the opinion of this author that Persephone points to human themes that are either experienced so deeply within a person or so broadly among the mass of humanity, that it is useless to apply such a technique to separate individuals whom we only know by their outer accomplishments (i.e., famous people). There is also a tendency for interpreting techniques to be applied to charts of people whose lives include circumstances that seem to superficially mimic the plot of the myth, as in the case of a famous kidnapping or a rape victim, with the emphasis placed on the mythic event rather than the awareness involved. We are trying instead to communicate the essence of the symbolism and the meaning or potential for growth of consciousness, leaving much of the practical application to the reader.

With that in mind, let us look at a few examples from our files:[174]

We see many examples where Ceres or Pluto play their part in the astrological drama with Persephone, as in the case of a woman with a natal conjunction of Saturn and Persephone who felt she never had successful relationships. With transiting Persephone conjunct her natal MC, she began a new career. With transiting Pluto forming into an exact square to Persephone and the IC, and simultaneously making a semisextile to her natal Mars, she came home one day to find her roommate dead. Persephone transits have a contradictory effect of both slow build-up and sudden change through separation-death.

We have the case of a woman whose natal Persephone is conjunct her Moon and square to Ceres. She spent most of her youth and early adulthood in attempts to separate from an extremely possessive mother, finally establishing her separate identity by having a child herself. When she became pregnant "out of wedlock," transiting Persephone was conjunct her natal Saturn, her mother disowned her and did not resume communication with her until she married seven years later.

The possessive mother image of the Persephone myth displays the extreme emotions centered around separation of mother-daughter at marriage time or the entry into womanhood. The mother who unconsciously attempts to keep the cord connected is acting out of fear of separation and general repression of the masculine principle. The smothering mother is portrayed clearly in the myths. Persephone was an out-of-wedlock, "love child," for Jupiter and Ceres were the parents of Persephone, but Jupiter rejected Ceres and married Juno instead. Ceres can, in a negative projection, represent the mother who, being rejected or ignored by a male partner-spouse, pours all her attention on the offspring. Ceres truly worshipped Persephone. Her attitude might have been something like: "So what if Juno gets Jupiter, then at least I'll have Persephone." Her protective possessiveness of her daughter reflects her lack of connection to (masculine) power. Women are apt to act this out as a fear of consciously grasping and manifesting the masculine principle. (Here again is the male-female, mind-feelings split.)

The Persephone myth can also reflect the healthy resolution of the feminine maturing process. It should be noted that each one of the "difficult" experiences related in this chapter were considered by those involved as avenues to greater consciousness. It is the law of growth that you can't make an omelet without breaking eggs.

Persephone does seem to be present in rape situations. We have the file of one woman who was raped when progressed Mars conjuncted her natal Persephone. Another case is a woman who experienced "date rape" the very day that transiting Persephone stationed at a position within 17 minutes of conjuncting her natal Pluto. Persephone is likely to be prominent in the charts of both rapists and raped women. We also note in this regard the extensive media coverage of mass raping of women in Bosnia during 1993, the year that Pluto and Persephone were locked in a continuous square at 22-to-24 degrees of Scorpio/Leo.

We have another example of a woman who experienced the following events during the 3-year period that transiting Persephone was within 1.5-degree orb of a sextile with her natal Uranus and a semisquare with her natal Ceres: She left her husband, taking her children to go live with her mother, then returned to her original (pre-marriage) hometown. During this period she migrated to and from seven different homes, and formed deep friendships with two very Plutonic men (one already married and one gay).

Persephone's influence is not limited to women. We have the case of a man who was divorced when his Persephone conjuncted his 7th house cusp by solar arc, from a wife whose progressed Persephone was inconjunct her Sun. Within a year the man remarried a woman with Persephone conjunct her Ascendant natally. Another man was experiencing his marriage as restrictive while transiting Persephone

Persephone Is Transpluto

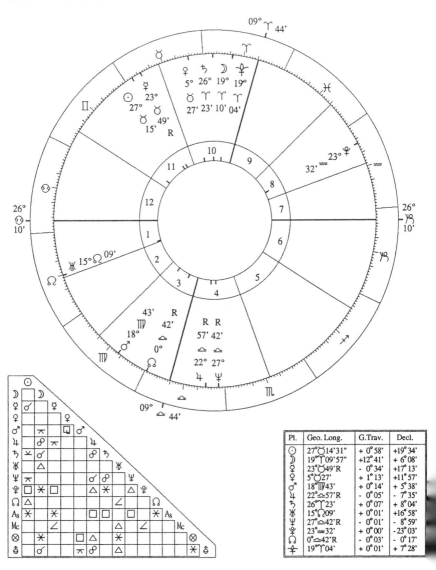

New York Stock Exchange	08:52:00 AM LMT	May 17, 1792	
New York, NY	42°N 43'00"	73°W 57'00"	
Placidus	0:39:58 S.T.	J.D. = 2375712.0776	
Natal Chart	R.A.M.S. = 3:43:46	Obl. = 23°27'59"	
Tropical	No Ayanamsha	Geocentric Ecliptic	

Pl.	Geo. Long.	G.Trav.	Decl.
☉	27°♉14'31"	+ 0° 58'	+19° 34'
☽	19°♈09'57"	+12° 41'	+ 6° 08'
☿	23°♉49'R	- 0° 34'	+17° 13'
♀	5°♉27'	+ 1° 13'	+11° 57'
♂	18°♍43'	+ 0° 14'	+ 5° 38'
♃	22°♎57'R	- 0° 05'	- 7° 35'
♄	26°♈23'	+ 0° 07'	+ 8° 04'
♅	15°♌09'	+ 0° 01'	+16° 58'
♆	27°♒42'R	- 0° 01'	- 8° 59'
♇	23°♒32'	+ 0° 00'	-23° 03'
☊	0°♎42'R	- 0° 03'	- 0° 17'
⚴	19°♈04'	+ 0° 01'	+ 7° 28'

conjuncted his natal Saturn. As transiting Persephone came to exact conjunction with his natal Ascendant, and Persephone by solar arc progression made a sesquiquadrate aspect to his natal Moon, the man sought a major redefinition of his relationship with his wife, with the hope that his marriage would not end in a traditional "messy" divorce.

Persephone's motion being so slow (only half a degree per year), we must check solar arc progressions in natal charts, but we also can use her for analysis of long cycles in astro-history. One manifestation of this can be seen when we analyze the chart for the "birth" of the New York Stock Exchange (see chart on page 124). Persephone is exactly conjunct the Moon, which would seem to emphasize the up-and-down (ascending and descending), cyclical nature of American economics.

Persephone's slow movement also means that entire generations share the same sign placement. Persephone has been located in the sign Leo since the late 1930s. She and Pluto entered Leo together in conjunction at the beginning of World War II. Everyone born between 1939 and 1957 have Pluto and Persephone in Leo in their charts. The parents of the Pluto-and-Persephone-in-Leo generation had Pluto and Persephone natally in Cancer. In general, we see this reflected as a "Me" (Leo) generation that needed to learn individuality via separation from the bond of traditional family (represented by the parents of the Cancer generation). The conjunction of Persephone-Pluto at the cusp of Cancer-Leo highlighted the separation of the twentieth century into pre- and post-War eras. It also extended the meaning of splitting what was thought to be a basic unit of matter, the atom, to the disintegration of what was thought to be a basic unit of society, the nuclear family.

Cancer is a sign with a tendency to identify with emotions and to fear losing possession of family. The classic freedom-closeness dilemma is seen in the signs which are inconjunct to Cancer; Sagittarius and Aquarius. Persephone situated in Cancer released unconscious potentials for the women who "mothered" what has emerged since W.W. II. The power center of the home was put to stress; nurturance was redefined (bottles and formula replaced the "nurse"); women began separating from their primary role as mothering homebody. Men had a parallel stress in dealing with the equality issues brought on by women entering the workforce.

The Persephone myth is clearly an image, both receptive and reflective, of relationship in all its variations — not just through male-female partnering, but also via parenting. And parenting involves the experience of children. We have focused on these ideas as they manifest in the "adult" world of marriage, politics, culture. Aside from such large and complex influences of social environment which we can analyze from our "adult" point of view, what about the microcosmic world of children?

ANOTHER "SLEEPING BEAUTY" AWAKENS

The sign Leo rules children, and with Persephone in that placement since World War II, it has been a time to listen to the children. They mirror society's repressed energy. The current issues of child abuse, teenage suicide, and milk carton kidnappings all reflect this stress. There are, on the one hand, the dominant cultural forces which are so pervasive and subtle (despite their brutality) that they are "picked up on" and reflected by children subconsciously. On the other hand, there are also the images (and myths) that are deliberately packaged and presented to children in a conscious manner. An examination of children's books (meaning those books which have been written for children to read), particularly those that present the Persephone myth, can tell us something about the larger cultural attitudes and how they are passed on to children.

We study myths and folktales, questioning their origins, looking at them from an historical perspective, watching how a particular age or culture (including our own) interprets a myth and alters it. We can make generalizations about what the myth means, and we can notice the exceptions and the variations. We realize that we might (for example) be reading or hearing a modern author's version of a story that was first written down (we think) in the Renaissance, but that version was actually a translation of another translation (and so on). We have to conclude that myths are not ends in themselves, but an on-going process ... a Neverending Story. What this means is that, asking ourselves whether a particular story is the original is a rather rhetorical question.

What is really being asked, as any child can tell us, is: (1) Is it a good telling? and (2) Is it full of meaning? Myths live on because they've got the basics of good style and relevant content. But they live on because of one other obvious reason: every time someone (anyone) tells the story, that myth lives on just a little longer.

As they say in the New Age 'creative imagination' workshops, *you create your own myth*. Likewise, from a socio-cultural standpoint, we are creating our own mythology. The folklorists of the future (down the road a few generations) may look back and comment that the folk tradition during the last half of the twentieth century was distorted beyond recognition by a particular kind of religious didacticism ... a religion whose god was not spiritual but material, whose effigies were not carved by artists, but by the Mass Media: "Buy this and you'll have eternal life, youth, sex; buy this and you'll live happily ever after." We all know that more children in this culture can rattle off advertising jingles than can remember Mother Goose rhymes. Traditional folk tales, orally transmitted, are indeed an endangered species.

Our attempt here to become discerning in the process of passing on folkmythic tradition is complicated. Picking our way through modern revised classics and Disney adaptations, we also have the concern about how new myths are being created to replace, enhance or distort older ones. (When E.T. phones home and his friends come to pick him up, he's not going to take his myth with him.) But this is not an argument about which story or version thereof is of greater value. There are also some inherent dangers in becoming too self-consciously critical in analysis of myth. Myths operate like dreams ... there's a point where we must just let the dream happen or else we fool ourselves into thinking we can control the flow, creating a contrived version of our own reality. And then our myths begin to sound like they're trying to make a point, losing all the spirit of fun and spontaneity, along with the subconscious alignment with flow. In other words, we have to trust that what is valid will endure. We have to leave the validation of new myth to the future that looks back on our present as its past.

Some stories will die out as the new forms and replacements are created. But if history repeats itself, some stories and myths may not be dying, but just going to rest for a while. Like Sleeping Beauty, they gather energy until some consciousness awakens. The Persephone myth is a sleeping beauty of sorts. Looking at children's literature as a measure of "how we pass on our myths," we can make some observations.

The story of the Earth Mother Goddess Demeter and her daughter Persephone lay relatively dormant and anonymous for thousands of years, content to be "just" a minor contribution to the various editions and renditions of the collected Greek Classical Myths.[175] Then, all of a sudden, in the space of three years (1971 to 1973), no fewer than four separate children's picture books were published that were devoted solely to this one particular story. Three of those were produced by major, award-winning children's book authors and illustrators: Barbara Cooney, Margaret Hodges,[176] and Penelope Farmer (well-known in Britain).

This is a demonstration of the Jungian concept of synchronicity, that a myth emerges on the common denominator level (children) just as the consciousness that it represents awakens. The early 1970s were years when the Women's Movement achieved new expression and power in the public eye, a reflection of the myth that tells of an archetypal relationship of dependence and of the struggle to grow out of that dependence towards autonomy as a woman. 1973 was the watershed year of the precedent-making (and currently under dispute) Abortion Law which concerned itself with definitions of life and death. And here we have a story about death and rebirth; telling us that death and separation are just part of the cycle that leads to new growth. It was also the early 70s when environmental and ecological awareness became a real public issue, and here to tell the story was a myth about the power of the Earth Goddess, telling us it's not nice to fool (with) Mother Nature.

The awakening during the early 1970s was indicated astrologically by a great deal of activity among the planets Persephone, Pluto and Ceres: During the years 1971 - 1973 Persephone was situated at 15 degrees (+/- 1 degree orb) Leo, one of the powerful midpoints of the cardinal cross, and in semi-square to Pluto, who stationed retrograde at 29 Virgo (an important position for Ceres), entered Persephone's sign of

Libra, and stationed again (direct) at 29 Virgo. Pluto's permanent ingress in August 1972 initiated twelve years of Pluto-in-Libra, meaning Pluto was "with Persephone" in the sign she rules. During this same period Ceres spent an unusually long time (9 months) in Leo (near Persephone), and most of the rest of the period in Virgo, Libra and Scorpio, the respective signs of rulership for these planets.

Examining the children's books telling the Persephone story gives us some fascinating clues to the society's inner and outer contradictions over the meaning of Persephone's myth. The text in Barbara Cooney's *Demeter and Persephone* (Doubleday, 1972) is not an adaptation but an actual translation of the Greek "Homeric Hymn to Demeter." This is relatively old as most stories go, but many sections of the Persephone myth pre-date Homer, and can be found in nearly identical form within the myths of Sumerian and Egyptian heritage back to about 3000 B.C.

Barbara Cooney's illustrations are delicate, nostalgic, sentimental, and correspond to the rather idealized and idyllic picture that many people mistakenly hold as the image of ancient Greece. The "favorite part" in this story for the children is what we might expect: Persephone returning to her mother. The dependence theme is emphasized here. The presentation shows no realistic contrast of this world against the terrifying unknown of the underworld and subconscious. While the text describes Persephone's abduction as her "shrieking shrilly," the illustration makes her look like she's going out with Pluto for a Sunday drive.

But let us turn to a different children's version, *Daughter of Earth* (Delacorte, 1984), produced by another award-winning author, Gerald McDermott, in the year that Pluto entered Scorpio.[177] McDermott has succeeded at portraying the power of this story with vivid, dynamic lines and colors. He shows the terror, the darkness, the unknown. Comparing McDermott's depiction against Cooney's "Sunday drive," we can unreservedly say "Now, this is a *Rape.*" The author has managed to make a book for children about the abduction and rape of that very state in which children exist: innocence. And he has told this story of growing-up in a way that allows the images of the myth to remain symbolic and not offensive to the touchy parents who wish to censor their children's exposure to real life. (The opposition of well-intentioned parents to the apparent violence of fairy-tales and myth is often a disguise of their fear of their own subconscious.)

Persephone's myth is about that powerful meeting place in each of us, the point-of-contact where the earth is ruptured, where our conscious, rational mind meets our subconscious (the underworld). When children see McDermott's version, they don't say that their "favorite part" is the tearful reunion; they are inspired instead by the scenes of risk, daring, and the unknown path that leads to independence ... when Persephone tugs at the flower and the earth opens up. In this version, it is Pluto (not Gaia as in the original tale) who has placed a Narcissus, the flower of egotism, as a snare with which to tempt Persephone. McDermott is very inaccurate with the original text on this point. What he has drawn is a daffodil, but in general he is true to the source he states that he has used, Ovid's *Metamorphoses*.[178] He also has shown details like the pigs, sacred to Demeter and Persephone. This is an exciting story and it deserves such bold drawings, although like most interpreters, McDemott makes Pluto out to be a creepy macho rapist, which he really isn't.

An interesting twist on the Persephone story is to be found in a Russian folktale version called the Snow Maiden.[179] A young goddess daughter ends up living with her husband during the winter on earth, but goes back during the warm months to her spiritual parents who live in other realms, because she is susceptible to melting. The Persephone myth is certainly versatile due to its inherent internal reversals. The world is literally turned upside-down, or inside-out, and seasons are turned around or stopped. But the story, as we know, is more than just an explanation for the change in seasons. If we read it with attention, we can discover an intensity and profundity which is equaled only by the depth to which we are willing to explore our own underworld, the story within ourselves.

PERSEPHONE AND PART-TIME RESIDENCY

Astrologers have used the charts of nations and world leaders to give macroscopic views of the world we live in. But "small is beautiful,"

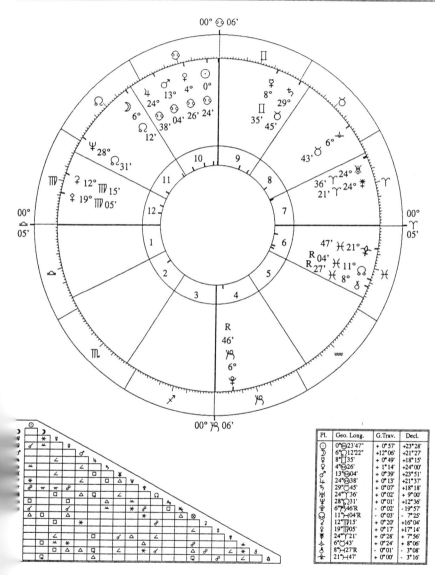

Town of Ashfield	12:00:00 PM LMT	June 21, 1765
Ashfield, MA	72°W 47'20"
Placidus	6:09:18 S.T.	J.D. = 2365885.2081
Natal Chart	R.A.M.S. = 6:00:28	Obl. = 23°28'11"
Tropical	No Ayanamsha	Geocentric Ecliptic

Pl.	Geo. Long.	G.Trav.	Decl.
☉	0°☊23'47"	+ 0°57'	+23°28'
☽	6°☊12'22"	+12°06'	+21°27'
☿	8°♊35'	+ 0°49'	+18°15'
♀	4°☊26'	+ 1°14'	+24°00'
♂	13°☊04'	+ 0°39'	+23°51'
♃	24°☊38'	+ 0°13'	+21°37'
♄	29°☊45'	+ 0°07'	+18°18'
♅	24°♈36'	+ 0°02'	+ 9°00'
♆	28°☊31'	+ 0°01'	+12°36'
♇	6°♑46'R	- 0°02'	-19°57'
☊	11°♍04'R	- 0°03'	- 7°25'
2	12°♍15'	+ 0°20'	+16°04'
♀	19°♍05'	+ 0°17'	+17°14'
♇	24°♈21'	+ 0°28'	+ 7°56'
♃	6°♉43'	+ 0°24'	+ 8°06'
♂	8°♑27'R	- 0°01'	- 3°08'
♃	21°♓47'	+ 0°00'	- 3°16'

131

and we can also gain perspective on local issues and our daily lives in the microcosm. Our everyday experiences can be affected as much by a local election as by a decimal drop in the international fluctuations of gold prices.

As an example, we show here the birthchart for a small town in Massachusetts. (See page 131) Since the exact "birth" time is unknown, noon has been used, a standard practice that places the Sun near the Midheaven as representing the town's personality being in the tenth house of public domain. We will first describe briefly the general character of the town, and then comment on the relevance of Persephone's placement in the chart.

We can see that the main focus (Sun), life blood (Mars), wealth (Jupiter), and values (Venus) of the town of Ashfield all spring from dairy farming (Cancer rules milk). Ashfield was an important source of Boston's butter supply during the 19th century, and the dairy farms are still a prominent feature today. With Pluto opposing all those Cancer planets in Capricorn, the town's "career" was always rooted (4th house) in sturdy hard work.

This attitude survives today in the crusty, cynical New England farmers who fight for their existence in competition with Agribusiness. In 1987 the milk surplus "crisis" (invented by the Agri-Dairy-Business) hit small farmers like those in Ashfield, and numerous cow herds were dissolved as Neptune transited through the 4th house of Ashfield's chart in early Capricorn. One farmer after another was forced to sell out. Hundreds of bossies disappeared, no doubt to find their Plutonian end and rebirth at some Burger King. But the Ashfield farmers are used to this kind of tough economics (Ashfield's Saturn is in Taurus).

Despite the agricultural emphasis and its small size, Ashfield has always been an intellectual center (Mercury in Gemini, 9th house), has been active as a place of publishing, and is the home of many writers. With Uranus in Aries, there are many inspired types working "solo." One famous master of illusion (Neptune in Leo) was born here — Cecil B. DeMille.

Ashfield was "born" on the Summer Solstice. With a noontime Sun, the four major directions of the chart (Ascendant-Descendant and MC-IC) are aligned with the solstice and equinox points. This makes sense because the turn of the seasons plays a major role in the town's life. The Annual Fall Festival, the biggest event and money-maker of the year, coincides with the transiting Sun's presence in early October in the first house, squaring Mars and Jupiter (there is always lots to eat and vigorous Morris Dancing at the festival).

The spring equinox is highlighted in Ashfield by Maple Sugaring, another big income producer. The hard work followed by excessive and jovial eating is again shown by the transiting seventh house Sun square natal Mars and Jupiter. The solar transits to the Midheaven and the Summer Solstice mark the arrival of summer residents who come to live in the Lake houses. The Winter Solstice transit of the Sun coincides with the Nadir point in the chart, and thus comes the proud claim by the locals that a *real resident* (4th house home-dweller) is one who endures the harsh winter weather.

A reader might ask, "But isn't this change of seasons true for everyone, not just Ashfield?" Yes, and so are all the attributes of the entire Zodiac present in each human being, regardless of birthdate. But if we look to the planet Persephone and a bit of Ashfield's history, we can enlarge upon the theme of part-time residency.

Persephone is positioned in the 6th house, stationary-retrograde at 21 Pisces 47. The only close aspect she makes with the major planets is a trine with tenth house Jupiter, connecting her with the wealth and expansion of the town. We recall from Persephone's myth that Jupiter played a lead role in determining that she would spend part of the year with Pluto and the rest with her mother Ceres on earth. There is a curious series of events in Ashfield's history that demonstrates this coming-and-going trademark of Persephone which links her with the town's economy.

Around the year 1812 an enterprising man settled in Ashfield, began raising peppermint, and soon had a distillery for converting the plant to peppermint oil. Persephone at that time was positioned at 6 Taurus, conjunct the natal Vesta, goddess of ritual fire and alchemy, associated with healing herbs used by women. Persephone was also trine Pluto (in ambitious Capricorn). Over the next six years, while Persephone moved into a sextile with Ashfield's Mars and a trine with natal Ceres, five such distilleries were created and the population of Ashfield reached an all-time peak of nearly two thousand souls. By 1830, with transiting Persephone sextiling natal Jupiter, there were ten places distilling peppermint, wintergreen, tansy, and hemlock, all important remedies for female "complaints," fertility control, birth and lactation conditions. The mint production is a very important factor. As we shall discover in a later chapter, mint appeared as a key ingredient in the Persephone myth.

Ashfield's "boom" was shortlived, for as transiting Persephone conjuncted natal Saturn, *another* enterprising man visited from N.Y. State, saw the profit in such a business, and returned to his settlement with peppermint roots from Ashfield, which he discovered could be cultivated much more easily in the milder climate of New York. As the Phelps, N.Y. distillery business began to flourish, so many peppermint-raisers and their families migrated from Ashfield that the population dropped drastically and never recovered. Historians of Ashfield recognize that the rise and fall of the peppermint industry directly affected the population.

Solar transits to natal Persephone give a more detailed picture of the seasonal changes in Ashfield's population. When the Sun passes each year into Ashfield's sixth house, the work and preparation of maple sugaring begins. As the Sun conjuncts Persephone and trines the tenth house Cancer stellium, the Sap Season reaches its peak, and Ashfield is host to hundreds of out-of-state visitors. The summer lakeside residents begin appearing after school is out and the children "return home," about the second week in June when transiting Sun squares Persephone at 21 Gemini. The Foliage Season begins about mid-September as the Sun opposes natal Persephone, bringing "leafer-peepers" from all over the country. As the Sun crosses the Ascendant at Fall Equinox, it "goes underground" (below the chart's horizon) like Persephone, and only the full-time residents remain in town.

As we hope to show with this example, Persephone is a good indicator of any populations that come and go with seasonal change. She might perhaps be a useful measurement for studying migrant workers, those that follow the harvest of various crops, or other groups of people who move between different worlds of experience. On a more esoteric plane, she can tell us about the movement between different levels of consciousness.

Astrologers have long interpreted many of the types of incidents related in this chapter as Plutonic. They have identified much of the issues of death-rebirth, sex, subconscious, and masses of people as ruled by Scorpio. With the emergence of Persephone from "beyond" or "behind" Pluto, we can refine this view and expand upon it. Death (or divorce, or separation, or loss) is not the end; there is something beyond it. As Persephone resolves her relationships (develops her Libran nature) with her mother-Ceres, her partner Pluto, and the Earth itself — as we all "work" these energies within ourselves — greater wholeness and balance is possible.

Chapter 13

THE MODERN ERA
AND PERSEPHONE IN CANCER/LEO
1882 TO 2011

Most astrology books that concern themselves with a single planet will include a "Grand Tour" of the Zodiac, describing the planet as it functions in each one of the signs, and giving interpretations for a variety of birthdates. Because Persephone's ephemeris is tentative and her cycle so long, we have chosen instead to focus on the only two signs which any living human can have for their birthchart Persephone position ... Cancer and Leo.

As "death-partner" to Persephone, Pluto provides a useful comparison for understanding the current position and movement of Persephone. Pluto is a slow-moving planet, distanced at close to 40 times as far from the Sun as our own Earth. He completes one turn of his orbit in 247 years, and spends about 20 or 30 years in each sign. Persephone has been calculated to be situated about twice that distance from the Sun, with a cycle of about 685 years. She spends from 40 to 80 years in each sign. The reason that both planets can vary in range of time spent in a sign is that, unlike other planets, neither travels on a steady path close to the ecliptic.

Observed from the Earth, Pluto moves through the Zodiac about 1.5 degrees per year (4 degrees direct and 2.5 degrees retrograde). Persephone moves even slower, direct about 1.5 degrees, retrograde about 1 degree, thus gaining a total forward movement of a little less than 1/2 degree per year. Pluto spends about the length of one generation's time in each sign; Persephone's stay in one sign is enough for three generations to be born, equivalent to the length of one human lifetime.

Consulting the ephemeris, we find that Pluto and Persephone have experienced quite an active relationship during the twentieth century. They were conjunct off-and-on in late Cancer during the late 1930s and entered Leo together at the beginning of W.W.II. Pluto has since that time moved through four Zodiacal signs (Leo, Virgo, Libra, Scorpio). Ever since Pluto's entry into Scorpio in 1984, the "Pluto-in-Leo Generation" (born 1939 to 1957) has been experiencing transiting Pluto square their natal Pluto.

But everyone born since 1939 has natal Persephone in Leo; she moves so slowly that she won't enter Virgo until 2011. The table below gives a very abbreviated listing of the Pluto and Persephone positions since 1939. It can be used to make some interesting observations about the Pluto-in-Leo generation by considering Persephone by transit.

APPROXIMATE POSITIONS OF PLUTO AND PERSEPHONE

These are approximations of the yearly positions when planets are stationary-direct. Actual positions vary by +/- 4 degrees.

	Persephone	Pluto		Persephone	Pluto
1939	0 Leo	29 Cancer	1977	16 Leo	11 Libra
1941	1 Leo	2 Leo	1979	17 Leo	17 Libra
1943	2 Leo	5 Leo	1982	18 Leo	24 Libra
1945	3 Leo	8 Leo	1984	19 Leo	0 Scorpio
1947	4 Leo	11 Leo	1987	20 Leo	7 Scorpio
1950	5 Leo	16 Leo	1990	21 Leo	15 Scorpio
1952	6 Leo	19 Leo	1992	22 Leo	22 Scorpio
1955	7 Leo	24 Leo	1995	23 Leo	28 Scorpio
1957	8 Leo	28 Leo	1997	24 Leo	3 Sag
1959	9 Leo	30 Leo	1999	25 Leo	10 Sag
1962	10 Leo	8 Virgo	2001	26 Leo	13 Sag
1964	11 Leo	12 Virgo			
1966	12 Leo	16 Virgo			
1969	13 Leo	22 Virgo			
1972	14 Leo	0 Libra			
1974	15 Leo	4 Libra			

The Baby Boomer generation (those who "grew up" to be the "flower children" of the late 1960s) were born with Pluto in the middle degrees of Leo, and experienced transiting Persephone conjunct their natal Pluto during the late 60s and early 70s. (A major portion of those who fought and died in the Viet Nam War, or who opposed it, had transiting Persephone conjunct their natal Pluto.) This was a time of the discovery of Dionysian energies (Make Love, Not War), mass migrations of Peace Marches and the Return to the Land movement, experimentation with new forms of family-partner-communal relationships. Homebirth and natural breastfeeding were re-discovered. This generation led the way into experimenting with consciousness-altering through drugs, mushrooms, herbs, meditation, and just "getting high on life." They decorated their "pads" (homes) as well as their bodies with organic designs, flowers, dayglow paint. Men wore long hair and were criticized for looking like women.

All these activities were manifestations of the awakening of the feminine aspect through Persephone. We can look to the myth of Persephone for illustrations of these themes. But it was also a time of the widening of the generation gap. Ceres, as the mothering principle, figures large in her daughter's myth, and experienced separation from her daughter. The mythological figures that form a complex centered around Persephone are Pluto, Ceres, and Dionysos. All are important elements interacting with Persephone in the stress and transformation of consciousness, and the re-definitions of gender, birth-death, and family-partner relationships.

The parents of the Pluto-in-Leo generation, as well as many of our current political leaders and other authorities or people in positions of power, all were born with Persephone and Pluto in Cancer. Theirs is the generation working through the Cancerian fear of loss of possession, which was accentuated by their experience of the Great Depression of the 1930s. (Persephone went stationary retrograde within two days of the famous stock market Panic Day of Oct. 24, 1929.) With an emphasis on Cancer, this generation's challenge centers around identification with personal emotions. The Persephone-Pluto conjunction in late Cancer and double ingress into Leo around 1939 released unconscious potentials which laid the foundation for many of the socio-economic ills of today.

This was especially true for women of this Persephone-and-Pluto-in-Cancer generation. Home to them was a center of power, but women were separating from their sole definition of self as mother-homebody. Men experienced a parallel stress as women sought careers and equality in the workplace. The women of this generation were the Rosie-the-Riveters who manned the factories and "held the fort" together at home while their men went off to war, but when the males came back as heroes, social pressures forced women back to traditional roles. Technology and the World Wars had, in reality, destroyed much of the basic family unit, but this generation continued to cling (with a tenacity only Cancers can have) to traditional ideals. Insecure Cancerian urges prompted tighter tying of apron strings as pressure toward recognizing freedom and individuality grew. Once they had raised their offspring, this Pluto/Persephone-in-Cancer generation was personally threatened by a fiery Pluto/Persephone-in-Leo generation whose women were burning bras (Cancer rules the breasts) and whose men were burning draft-cards (symbol of obedience to the Mother-Cancer-country).

The older Cancerian generation had a personal connection to the national identity, symbolized by the Cancer Sun and stellium in the birthchart of the United States. Patriotism was considered a cardinal rule to this Cardinal-sign generation. So identified were they with nationalism that they experienced the Leo generation's opposition to the Viet Nam War as a personal affront. Although there are many "reasons" for the discord in this country during the late 60s-early 70s, we can attribute much to the generational differences signified by Persephone-Pluto.

With Pluto's entry into Scorpio, squaring the Boomer generation's natal Persephone, we have observed the increase in single-mother population. Although widely recognized as a problem, it was denied and/or criticized by former President Bush (one of the Cancer generation), whose natal Persephone made a T-square with his Moon-Saturn in Libra and Ceres in Aries. During Bush's administration, "Family Values" were raised (or rather, *lowered*) to the status of political raw-material to be hashed and homogenized during the 1988 election by politicians who were out-of-touch (astrologically and otherwise) with current realities.

Persephone Is Transpluto

Barry Lyne's US Chart	04:47:00 PM EST	July 4, 1776
Philadelphia, PA	39°N 57'00"	75°W 08'00"
Placidus	11:40:17 S.T.	J.D. = 2369916.4074
Natal Chart	R.A.M.S. = 6:53:49	Obl. = 23°28'06"
Tropical	No Ayanamsha	Geocentric Ecliptic

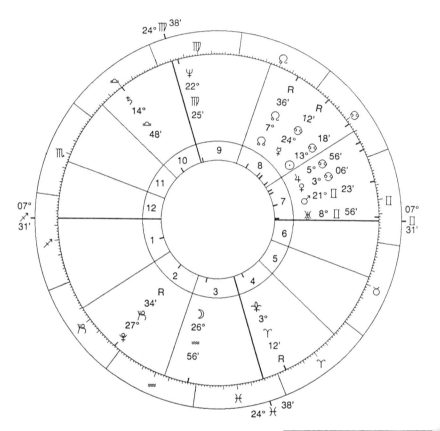

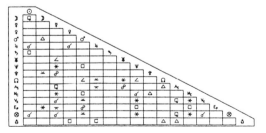

Pl.	Geo. Long.	G.Trav.	Decl.
☉	13°♋18'25"	+ 0° 57'	+22° 48'
☽	26°♒55'52"	+14° 32'	- 14° 16'
☿	24°♋12'R	- 0° 27'	+17° 29'
♀	3°♋06'	+ 1° 14'	+23° 32'
♂	21°♊23'	+ 0° 41'	+23° 34'
♃	5°♋56'	+ 0° 14'	+23° 16'
♄	14°♎48'	+ 0° 02'	- 3° 31'
♅	8°♊56'	+ 0° 03'	+21° 45'
♆	22°♍25'	+ 0° 01'	+ 4° 10'
♇	27°♑34'R	- 0° 01'	- 23° 44'
☊	7°♌36'R	- 0° 03'	+18° 24'
⚷	3°♈12'R	- 0° 00'	+ 1° 17'

140

The years 1991 through 1993 witnessed a square of Persephone-Pluto at around 21 to 23 degrees of Leo-Scorpio. Readers have only to review their own personal lives during this period to remember that this was a period of stress of redefining relationships, values, and all expressions of separation/reunion ... a time of *transformation or else!* (for everyone, not just those with "sensitive points" being aspected).

It is interesting to note that only one other time in U.S. history did Pluto and Persephone enter a new sign simultaneously ... around 1882 Persephone had just entered Cancer, and Pluto made his first ingress into Gemini. And, not surprisingly, this year is recognized as a turning point in history as well as a borderline between another pair of generations well-known for their "gap." This time it was a generation with Pluto in Taurus and Persephone in Gemini who clashed with a generation born with Pluto in Gemini and Persephone in Cancer. The watershed period dividing their births was the early 1880s, and the time of their open opposition was, like our 20th century Cancer/Leo version, about 30 years following the double ingress, and just when transiting Pluto entered Cancer (early 1910s).

Historians Neil Howe and William Strauss have documented these patterns of generation gap in a book entitled *Generations: The History of America's Future* (1991).They noted that the pre-1882 group, which they call the "Missionary Generation," foreshadowed the experience of the 20th century "Boomer Generation." These earlier peers were raised, like the Baby Boomers, in the aftermath of a great War (their fathers were Civil War heroes), and were "indulged as children, spectacular as students, furious with soul-dead fathers, absorbed with the 'inner life,' unyielding as reformers, and slow to form families." They grew up in a "world of orderly families and accelerating prosperity, ... sickened with advantages," but upon coming of age, they rebelled against the Horatio Alger materialism, instigated anarchy and radicalism.[180]

Looking to the chart of the United States, we may surmise that Persephone will always be a "sore spot" for the American Mom-and-Apple Pie mentality. In 1776, Persephone went stationary retrograde on July 4th, at 3 degrees of Aries, in exact square to the Venus-Jupiter conjunction in Cancer. (See chart, page 140) This signifies a strong and independent feminine challenge to expansive materialism and the popular American female image of big-tit (Jupiter-Venus), dependent Mama (Cancer).

It is a widely-accepted tenet of astrology that the outer trans-personal planets, in addition to being determinants of complex and transformative human development, signify larger groups or mass levels of humanity on a collective or generational level.[181] Uranus, Neptune, and Pluto have been shown via their ingresses into signs and their interaspects to be identifiers of specific generations or periods in history. Persephone-Transpluto is no exception. We can look at the 20th century from the standpoint of her transits and see how she has affected all generations. We refer to the table above, noticing in particular the movement of Persephone through Leo, along with the years when Pluto passed through Leo and Scorpio. We compare this with the following list, observing the almost continuous stress by way of transiting hard aspects that Persephone and/or Pluto make to birthcharts and generations with outer planets in the fixed signs.

(Years are approximate; check ephemerides for positions on specific dates)

1912-1920	Uranus in Aquarius
1916-1929	Neptune in Leo
1934-1942	Uranus in Taurus
1938-1957	Pluto in Leo
1955-1962	Uranus in Leo
1956-1970	Neptune in Scorpio
1974-1981	Uranus in Scorpio

What this list shows is that nearly everyone born in the past 80 years has experienced major hard aspects to their natal outer planets by transiting Persephone-Pluto-in-Leo or Scorpio. Even during the gaps when no known outer planet was in a fixed sign, Persephone was within orb of conjunction with Pluto (early 1930s) or a semisquare (early 1970s). Natally or by transit, mass numbers of humans alive today have had or are having to deal with serious Persephonic challenges. Persephone in Leo offers all of us, not just the Pluto-Persephone-in-Leo generation, the opportunity to heighten our understanding of this sign — a necessary step we must take before we can move across into the Aquarian Age.

Chapter 14

SCIENCE, MYTH, ASTROLOGY, AND HISTORY: THE PERSEPHONE CONNECTION

Persephone: Previous chapters have discussed WHO she is (her name and myth), HOW she expresses (her rulership of Libra), and WHERE she is (currently in Leo). The next question to be addressed is WHEN (how long is her cycle of one turn through the Zodiac?).

In the strictly physical sense, astronomers and astrologers are still working on the determination of her exact orbit and cycle, but metaphysically she is present and here with us. Somewhere between these two polarities of science and intuition we can use astrological guideposts in a unified search for Persephone, both here within our consciousness and "out there" in the sky.We begin by approaching this "unknown" planet with the same methods already familiar to astrologers in working with the "known" planets. We have historical evidence to correlate with scientific evidence to postulate a probable orbital cycle.

We have seen from a previous chapter on astronomical evidence that during the past 150 years several astronomers have been calculating the cycle of Transpluto to be somewhere between 600 and 800 years. The Titius-Bode Law (see table below), used for determining planetary distances, gives a value of 678 years. The tentative ephemeris for Persephone published by Hawkins and others uses a cycle of 685 years. We can apply this cycle or *block of time* in the same way that we use the Lunar or Saturn cycles for understanding general phases of human development, regardless of sign position or specific delineation.

What stands out immediately with Persephone's cycle is its great length; nearly 700 years is a large chunk of time, enough for 30

Bode's Law

$$\frac{A + B2^n}{10}$$

where
A = 4
B = 3
n = - ∞, 0, 1, 2, ...

								ACTUAL DISTANCE	SYNODIC PERIOD
4 + (	0	•	3)	:	10	=	0.4	Mercury	0.39 a.u.	88 d
4 + (	1	•	3)	:	10	=	0.7	Venus	0.72	224.7
4 + (	2	•	3)	:	10	=	1.0	Earth	1.00	365.25
4 + (	4	•	3)	:	10	=	1.6	Mars	1.52	687
4 + (	8	•	3)	:	10	=	2.8	Ceres	2.77	4.6 y
4 + (	16	•	3)	:	10	=	5.2	Jupiter	5.20	11.9
4 + (	32	•	3)	:	10	=	10.0	Saturn	9.54	29.5
4 + (	64	•	3)	:	10	=	19.6	Uranus	19.19	84
4 + (	128	•	3)	:	10	=	38.8	Pluto	39.52	248
4 + (	256	•	3)	:	10	=	77.2	Persephone	77	678
4 + (	512	•	3)	:	10	=	154.0	?		

Bode's law includes the asteroids and gives the distance for Pluto rather than Neptune.

generations to be born, ten lifetimes, or 25 Saturn cycles. When we observe the cycles of inner planets (from a month for the Moon to 28 years for Saturn), we are looking at very human or individual-personal quantities of time ... periods that can occur within one individual's lifetime. With the cycles of the trans-Saturnian planets, we move beyond the scope of one person's experience, towards quantities that measure generations, eras, and historical periods (84 years for Uranus, 165 years for Neptune, a quarter of a millennium for Pluto).

With the period of Persephone, we have the opportunity to observe much longer cyclical patterns that occur in cultural and even geologic history. The following discussion is based on the astrological principle that we can use a given *cycle* (regardless of planetary position) to see *patterns* and thereby gain a perspective on history. In this case, the perspective comes via the meaning and interpretation of Persephone. The reader is, as usual, advised to hang loose with regard to traditional distinctions between fact and fantasy (history-science vs. myth-imagination), remembering that Persephone is about crossing borders.

685 years is a large enough amount of time to suggest a possible link with that very greatest of astrological cycles, the Great Year — 25,800 years of the Precession of the Equinoxes. Astrologers have divided the Great Year into 12 Ages of about 2150 years each. Persephone's cycle works out to be very close to one-third of an Age (more specifically, dividing 2150 years by *pi* — 3.1416 — gives 684 years). Using *pi* is not such an arbitrary invention, for the equation determining the cycle or circumference of a circle is *pi* times the diameter, which is the horizon, ruled by Persephone. (See figure below.)

circumference = cycle

circumference = diameter X pi
cycle = horizon X pi
Age = Persephone X pi

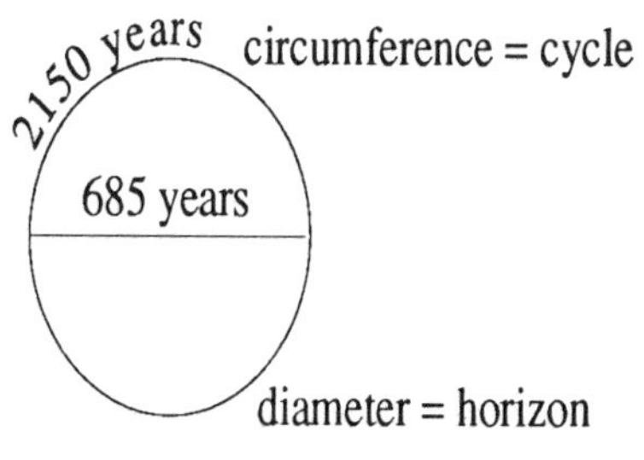

2150 years

685 years

diameter = horizon

We must start with our realization that Persephone's myth is about cycles (of separation and reunion) and is concerned with the very process of cycling itself. The resolution of her myth is that she will always spend part of the year sharing rulership of the Underworld with Pluto, and part of the year with her mother Ceres-Demeter on earth. Reviewing Persephone's myth once again, we recall the way in which this co-ruling arrangement came about:Persephone was abducted by Pluto and taken to the nether world. Her Earth Mother mourned her absence and consequently all plant and food growth ceased. Previous to this there had been a Garden-of-Eden existence of constant bloom and fruition, but then the earth experienced its first winter due to Ceres' grief at the loss of her cherished daughter.

Conditions on earth became life-threatening. The gods were worried that, if humanity perished, there would be no one to worship them, and then the gods themselves would die. After much diplomatic haggling on Olympus, Pluto agreed to release his new bride, but alas! Persephone had partaken of the food of the Dead and thus sealed her fate to be forever linked with Hades. She had ingested a certain number of pomegranate seeds (symbolic of fertility). It was finally negotiated on Olympus that the number of seeds Persephone had consumed would determine how she would divide her time between the two worlds (that of her husband and her mother).

The actual number of seeds differs in the various adaptations of the myth, and it is possible that this difference might have represented the varying length of winter or growing season in different locations (after all, the myth was widespread throughout the ancient world). However, a commonly accepted version says that Persephone ate three seeds and therefore must spend one-third of the year with Pluto. Because the Persephone myth has been known as a folklore explanation for "why we have the seasons," this has been usually interpreted to mean one-third of our human year, or four months of winter. Certainly this corresponds with the Northern Hemisphere's four winter month-signs (Sagittarius through Pisces), "guarded" at the beginning by Pluto's Scorpio and at the end by Mars-ruled Aries. The Sun signs Taurus through Libra likewise border the growing season. (See figure below)

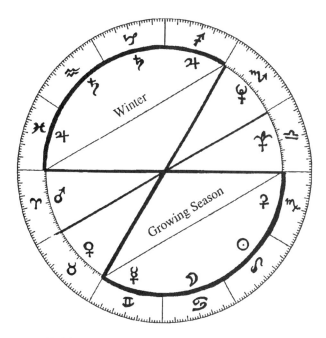

However well this seasonal explanation works, we should be on the lookout for more elaborate yet hidden astronomical data contained in Persephone's myth. It is certainly not uncommon for myths of various ancient cultures to "behave" like this. Stories such as the Mayan *Popul Vuh,* which tells of a Descent into Xibalba (the Underworld, pronounced Shibalba), have woven within them the mathematical and calendrical formulas for determining astronomical cycles of planets, eclipses, the Metonic cycle, etc.[182]

It is possible that this "one-third" quantity could be referring to a portion of the *Gods' Year,* rather than the human one of 365 days. Such analogies were common among the ancients who understood the Precessional Cycle as the mythical "Year of the Gods."

"To the deities a man's life seems like a day," said Herodotus (5th c. B.C. Greece), explaining that there are 70 years (about a human lifetime) for each "Gods' day" of the Precessional Great Year. In other words, there is a mathematical relationship between the God's Year

Division of Precessional Ages into Approximate Thirds		Concurrent Conjunctions of Outer Planets[185]	
(Cycles within the Precessional Ages of about 685 Years)			
approximate dates	exact dates (685 yrs. apart)		
4200 B.C. Age of Taurus begins	4166 B.C.	4180 B.C.	Uranus/Neptune
3500 B.C.	3481 B.C.	3492 B.C.	Uranus/Neptune
		3554 B.C.	Neptune/Pluto
2900-2800 B.C.	2796 B.C.	2808 B.C.	Uranus/Neptune
2100 B.C. Age of Aries begins	2111 B.C.	2122 B.C.	Uranus/Neptune
		2066 B.C.	Neptune/Pluto
14th-15th c. B.C.	1426 B.C.	1435 B.C.	Uranus/Neptune
8th c. B.C.	741 B.C.	747 B.C.	Uranus/Neptune
1st c. B.C. Age of Pisces begins	55 B.C.	84 B.C.	Neptune/Pluto
		62 B.C.	Uranus/Neptune
7th-8th c. A.D.	641 A.D.	623 A.D.	Uranus/Neptune
13th-14th c. A.D.	1326 A.D.	1307 A.D.	Uranus/Neptune
20th-21st c. A.D. Age of Aquarius begins	2011 A.D.	1993 A.D.	Uranus/Neptune

and that of the mortals. 25,800 divided by 365 equals 70(+) human years for each "day" of the gods.[183] Plato made a similar analogy in *Timaios* and in *The Republic*.[184]

Using our Persephonic "myth-amatical" formula, one-third of the Gods' Year would be 8600 years, which equals 12 times Persephone's cycle. So *Persephone's cycle is equal to about one-third of a Precessional Age* (rather than the whole Precessional-Gods' Year), and we may surmise that this might be one interpretation intended in the myth.

If we use Persephone's cycle to divide each Zodiacal Age into approximate thirds, we come up with some very suggestive dates (See table p. 148). The watershed years (which can only be approximate) happen to mark the transitions periods of large-scale cultural change. These years are highlighted by changes in climate, intensified earthquake-volcanic activity, subsequent mass migrations of entire races or tribes, and eventual re-synthesizing of cultures to incorporate foreign (i.e. alien) elements and to adapt agriculture (the human-Nature relationship) to new climatic conditions.

While referring to the above table, the reader may consider the following historical and scientific evidence, keeping in mind that we are observing a very slow-moving planet and broad, approximate time periods that follow a sequence of seven centuries.

Several old settlements in the Aegean were destroyed around 2100 B.C., and new ones emerged. Among the new ones were the recently discovered site of Mashkan-shapir, first settled around 2050 B.C.,[186] and the great Minoan Crete culture, which lasted until the time of the volcanic eruption of Thera-Santorini and associated earthquakes (15th c. B.C.). Matriarchal culture, centered in Crete and symbolized by the union of Persephone-Demeter, was torn apart by Plutonian (volcanic) influences. Large groups of Aryans were migrating into the Mediterranean during the 15th-14th c. B.C., and there is much evidence to show that climate patterns were changing at this time, causing once fertile areas to become dryer and cooler. The male deities of these Aryan peoples were usually associated with the worship of volcanoes.[187]

149

One historian has argued that the fall of the Mycenaen civilization (around the 14th c. B.C.) was due not so much to "an invasion from the outside, but an evacuation from within ... a dispersal."[188] He blamed drought and flood, quoting Plato, who said in *Timaios* that "fire and water" were the cause of the downfall.

The story was similar in Egypt. History of climate shows that around 2150 B.C. there was severe famine in Egypt due to low rainfall.[189] The great Middle Kingdom began around that time, with Egypt reaching a height of its power in the 15th-14th c. B.C. This was also the time that historians believe Moses and his people began their travels, initiating several thousand years of migrating Hebrew culture. It has often been conjectured that one event in the story of Moses, the famous parting of the sea, was associated with or actually caused by a major geologic event. Incidentally, some historians have dated another Biblical catastrophe, the Flood of the Noah story, as happening about one Persephone cycle earlier — 2100 B.C.

The history of Mesopotamia at this time shows a similar pattern. Ancient Babylonian records reveal that weather could be blamed for dramatic changes in harvest dates. Prior to the 15th c. B.C. the harvest occurred regularly in March. A change occurred after the 8th c. B.C. and the Fall of Nineveh, when the harvests regularly came a month later.[190]

The great civilization of Indus was flourishing until a rapid decline in the 15th-14th c. B.C. Failing monsoon rains have been suggested as a significant contributing factor.[191]

Mass migrations are another Persephone signature. Turning to Greece and the next phase or cycle division, Homer described the "wandering peoples" of the 8th c. B.C. as the "race of iron," whose arrival ended the Greek dark ages and began the Greek colonization of the Mediterranean. It is interesting to note that the middle third of both the Age of Pisces (7th-14th c. A.D.) and the Age of Aries (14th-7th c. B.C.) have been considered "dark ages" by historians.

Rome was founded in 753 B.C., and the culmination of the Roman Empire occurred about 700 years later. Its spread over the entire

European continent and much of North Africa and the Mideast was nearly complete by the time of Julius Caesar, circa 45 B.C. The first century B.C. was another watershed period, in the sense that the religion that would become Christianity was developing rapidly.

The 7th/8th c. A.D. period marked another shift as Christian Europe began a long period of holding off the invasion (attempted migration) of Islamic culture. Islam invaded Spain in 715 A.D. The 14th c. A.D. marked the final decline of the European Medieval ("Dark") Age, widespread Plague, the beginning of the Renaissance, the rise of the Ottoman Empire (which lasted over 600 years), and Marco Polo, who initiated centuries of European expansion and migration throughout the globe.

In the Americas, the early 14th century A.D. witnessed the end of the Anasazi Pueblo Indian culture as cave dwellings in what is now southwestern U.S. were abandoned. Archaeologists have explained the sudden departure as due to a severe drought period at the end of the 13th century, which may have been accompanied by intense geophysical disturbance. Sudden mass immigration had introduced this culture around 700 A.D.; mass emigration ended it about one Persephone cycle later.[192]

WHETHER PREDICTIONS

To summarize the history of changes in climate gives a perspective on the development of culture. There was a general warming period just after 3000 B.C. that resulted in a benign climate that encouraged the development of agriculture to spread north and west across Europe from Mideast origins. The rise of the Aegean, Egyptian, and Mesopotamian civilizations during the 3rd millennium B.C. (around our watershed of 2800 B.C.) coincided with greater water supplies in what are now dry zones.[193]

Climate historians recognize that the effects of climatic change in both the 15th and 8th centuries B.C. were widespread and included migrations, invasions, and upheavals in culture.[194] A general cooling spell began around the 15th-14th c. B.C., then a warming spell after

151

the 8th century B.C., then a generally dry and warm period with many droughts until around 700 A.D. when another cool/wet period began.[195]

Halfway through this next 700-year cycle, an intermediate warming spell occurred around the 10th-11th centuries A.D. Known as the "Medieval Climatic Optimum," it encouraged the Viking expansion into areas such as Greenland, which then had to be abandoned as the cooling spell resumed.[196] An erratic cooling spell began in the 14th c. A.D. which resulted in much crop failure and consequent widespread famine in Europe (1315 A.D.), just in time for the Plague.[197] Statistical studies show a cyclic pattern of drought every 170 years (one quarter of a Persephone cycle), as shown in the following figure:[198]

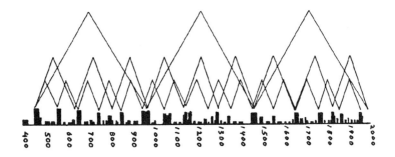

Cycles of Drought: 100 year, 170 year, and 510 year
from 400 to 2000 A.D.

There is much documentation for the linking of cultural shifts to changing weather and agricultural patterns, but what, in turn, "caused" the changing weather? Volcanic eruptions are considered a significant factor. Although scientific debate continues over the extent of influence on the climate,[199] the connection between volcanoes and climate is recognized. Volcanoes throw dust veils into the atmosphere which dim the sunlight, causing temperatures to drop, then summer crops fail, famine occurs, and migrations follow this chain reaction as people search for more hospitable environments.

Volcanoes and climate have an interchangeable relationship. Volcanic activity affects the weather, and weather evidently affects volcanoes. This is because warming trends can cause shifts in the amount of ice in glaciers and at the poles, and this does more than raise sea level. The weight of water in ice that covers land is released and moves to the oceans. Some climatologists suggest that this can set off volcanoes because the land is relieved of the weight of ice. Shifts of weight within the earth's crust trigger volcanic eruptions which in turn block the sun, producing the formation of more ice.[200]

From an astrological standpoint, the symbols involved in this historical-climatological ecosystem are Pluto (volcanoes), Ceres (agriculture), and Persephone (migrations). This interpretation is supported by the ancient Greeks. They regarded the equinoctal storms (which we experience in the Eastern U.S. as fall hurricanes and March blizzards) as the anger of Demeter over the abduction of Persephone by Pluto and her impatience for the delayed return of her daughter.

Using astronomical theory to explain changing climate and ice ages is not a new idea. Scientists have observed since the 19th century that global changes in climate are induced by subtle changes in the earth's orbit, which are in turn related to the long cycle of the Precession of the Equinoxes. This is now known as the Milankovitch theory.[201]

Considering the current global weather changes, the droughts and famines, the increase in earthquake-volcanic activity, and the great numbers of displaced populations, the late 20th century seems to be right on schedule with our proposed Persephone cycle. Changes in atmospheric conditions are sending entire animal and plant species into migration, there are multitudes of endangered species, disappearing rainforests, wetlands, and other ecosystems, polluted atmosphere, depleted soil, the "homeless," and unprecedented growth in refugee populations. Regarding the evident increase in earthquake activity, it has become apparent that, although improved technology for recording earthquakes accounts for some increase, the current figures are still striking. This is especially noticeable in the death tolls for the large "killer" quakes. Twenty of these during the past 40 years have resulted in over one million deaths. The number of deaths due to all types of earthquake in the year 1990 alone was nearly equal to the total for the entire decade of the 1980s.[202]

153

During the past decade there has been an interesting shift in the views of ecologists about how the structure of communities in Nature are formed. The old "separatist" theory of competition between the species has been gradually replaced by the more "wholistic" theory of the importance of climate.[203] There is also increased awareness as to the intruding role that humans and their technology play in the drama of geophysical events. Compared with premeditated human assaults on the earth's land, air, and water, the damage from earthquakes and floods is minimal.[204]

Astrologers have traditionally assigned rulership of volcanoes, earthquakes, and masses of humanity to Pluto; the rulership of nature, agriculture, and ecology-environmental issues to Ceres. With the interconnection of global weather patterns with geophysical disruption and consequent changes in agricultural production, ecological imbalances, and population movements, it would seem that Persephone is a linking factor. This is totally in keeping with the Persephone myth. It was the Earth Goddess Gaia who opened a passageway in the ground to allow Pluto up from the Underworld to perform his abduction. Ceres, of course, reacted by producing famine and then wandered herself, homeless in her state of mourning. Jupiter, considered the god of weather by the ancient Greeks, was the one who arranged Persephone's abduction and negotiated her release from the Underworld.

Persephone does not appear alone in her myth, nor in our reality, but rather *is always seen in relation and interaction with others.* We cannot isolate her influence from the full cast of interacting characters in her myth, and we should likewise not view geophysical disruption (Pluto), weather (Jupiter), agriculture (Ceres) and migrations (Persephone) as totally separate from each other. The scientific and historical evidence seems to show that, with long-term, global cycles, the Persephone Connection is worth investigating as an important astrological factor.

Chapter 15

THE POWERFUL FEMININE PRINCIPLE: SEXUALITY, EARTHQUAKES, RECYCLING, AND UFOs

Persephone is not *the* Goddess. There is no one name or single identity for "The Goddess," anymore than there is one universal name or single concept of "God." But in her myth, Persephone shares company with some of the most powerful and transformative figures in mythology, and by association she interconnects all of them. Pluto is of course the powerful lord of the subterranean world, Demeter and Gaia are both ancient Earth mother figures. They combine together with Persephone to represent our home planet, our foundation and the geophysical forces underneath. Their inter-relationship is reflected in Persephone's apparent astrological involvement with geologic phenomena such as earthquakes, volcanic eruptions, floods, and other major upheavals that mark the migration of earth matter between the world above, below, and across the horizon. We see her theme of separation-reunion as an analogy on a geologic level: When the earth is formed, there is the cycle of erosion, sedimentation, metamorphosism, erosion again. The Earth itself is unendingly broken down, separated, and reunited.

We can look to Persephone as a relevant character in the "earthquakes" of our lives, as well as in our attitudes about humanity's relationship with the earth (recycling versus pollution), and the potential relationship of humanoids with other populations, both the plants and animals of our planet and the aliens of the universe. Since the Earth has always been considered a feminine entity, and the sky (outer space) is considered "masculine," we conclude that *extra-terrestrial contact with humans is a macrocosmic projection of our own human male-female relationships.* And in order to know the Earth and ourselves, we must first come to an understanding about what exactly is "The Feminine."

155

THE FEMININE PRINCIPLE IN ASTROLOGY

In most astrological literature, Venus and the Moon are taken for granted as being "the generalized feminine" and as representing *The Goddess* (and women in general). The asteroids have been treated as separate parts of the fragmented "feminine," and in fact have been so disassociated that some have been "masculated" or even "neutralized" (as in the case of those being named for astronomers and cities). In approaching mythology from a feminine standpoint we are challenged to avoid the apparently common tendency to make broad generalizations about "The Goddess," or to present her multivariant aspects as fragmented. There is some truth to be gained from both methods, as long as we consciously recognize what we are doing, unifying or dividing. But Astrology has not approached this problem as deeply as it could.

As mentioned previously, astrologers have made many of their conclusions based on popularized, patriarchal versions of the myths or based on fashionable but superficial "feminist" interpretations. In either case, early and primary sources have not been carefully examined. This is lazy scholarship. We waste time re-inventing the wheel and we limit ourselves to accepting second-hand information when we fail to analyze what has already been deeply researched in established academic fields. We need to recall that Astrology was the parent to many modern-day disciplines —- psychology, mythology, anthropology, archaeology, history, and science. Just because these offspring have grown up to be different and rejecting of the "parent," that is no reason for Astrology to ignore their accomplishments and resist learning from them. It would be highly beneficial for astrologers to consider broadening and deepening their education and professional stature by studying areas outside their own field. Astrology will remain a second-class discipline as long as we astrologers insist on operating as if we were independent of consensus reality.

Throughout this book we have attempted to determine the identity of Persephone and to distinguish her from over-generalized, fragmented, and superficial definitions of the Feminine, and well as to use her to build an inclusive, expanded concept of the Feminine, the Masculine, the Androgynous, and Conscious Relationship. We can further this understanding by continuing to compare the research on Persephone with what is known about Venus, thereby refining our astrological principles about any of the female planets.

PERSEPHONE IS LUNAR, VENUS IS SOLAR

One of the most obvious differences between Venus and Persephone is found in their soli-lunar identities. The planet and mythological character of Venus has been almost universally associated with the Sun. The Mesoamerican world saw Venus as the "twin" of the Sun. Most cultures have traditions describing Venus as a "star blazing like the Sun," and this story was the same in ancient Greece. Of all the Greek goddesses, only Aphrodite was intrinsically "golden." While others were labelled "silver" (like the Moon), "golden" was the most frequent epithet of Venus, and it was she who more than any other goddess was unambiguously solar.[205] In her myths she seduced lovers by daylight; her sexuality was exposed and open; she expressed no ambivalence about sex; she was often portrayed nude. Aphrodite developed along the lines of the masculine principle of the Sun, yet she became the most "feminine" — in the conventional sense of patriarchal Greek and modern sexist culture.

The character of Venus as patroness of courtesans, as the talented and cultivated seductress (receptive yet actively wooing), and as the mistress of "wiles," all sounds suspiciously like a male projection of Woman as perfect sexual object. To cap off this masculine ideal, in Greek myth *Venus is never raped* ... a most unusual circumstance among the Greek pantheon. She also is rarely given negative attributes by mytho-psychologists; she does not slay her children or cause famine; her worst trait seems to be "subjective cruelty." An entire book may be devoted to Freudian analysis of the role in Greek myth of various destructive, devouring, entwining, castrating, and otherwise threatening females, but Aphrodite will hardly be mentioned.[206]

One of the most frequent epithets of Aphrodite is *philommeides*, which ostensibly means "smile-loving." Like women in toothpaste ads or the ever-grinning airline hostess, receptionist, or Miss America contestant — there's something just a bit too bright about this solar female. Is there never a cloud to darken her sunny *have-a-nice-day?* Why does she have no ups-and-downs, no range of emotion? Doesn't she ever have PMS? How could she, if she has no lunar nature?

A little etymology might be useful at this point. The term *philommeides* has been translated as smile- or laughter-loving, and is usually found in myth "in contexts which explicitly (or implicitly) emphasize Aphrodite's aspect as a goddess of sexual love,"[207] i. e., when she's on the make. But a second translation of this word is "loving a man's penis (or genitalia)," and yet another is "to own or possess male genitalia." Hesiod wrote in the *Theogony* that Venus "was called philommedea because she sprang from the members." We can view these multiple meanings as the *double-entendre* so typical of Greek poetry, but we can also see them as further confirmation that Venus-Aphrodite was created by, with, and from a patriarchal projection of masculine desire.

While the drives of sexuality are usually associated with nature, wildness, and therefore the Lunar aspect, Venus "elevated" this activity to a more solar realm of cultured and sophisticated love-making, as with the Japanese *geisha*. No matter how it's dressed up (Venus rules cosmetics, perfumes, adornment, and fashionable clothing), however, the underlying message is purely biochemical. Astrological Venus certainly does not denote any potential female qualities such as genius, athletic ability, or spiritual orientation. Her purpose is infatuation, which some anthropologists have said is a necessary preliminary to the survival (procreation) of the human species.[208] Her method involves the use of pheromones and primitive forms of social interaction. Her realm is unevolved and unconscious emotional reaction, which is why she's associated with attraction, attachment, jealousy, romance, and infidelity. This is not to deny the usefulness of Venus or to ignore the reality of mating behaviors that humans share with animals; the Venus level can certainly be fun. The mistake comes in limiting our experience of relating to this level, and in identifying "female relationships in the life of the querent" as simply Venus in the chart.

As we have explained in earlier chapters, Persephone represents a relationship with the feminine which is beyond the limited patriarchal-sexist view of women (symbolized by the solar concept of Venus). But how can we say with confidence that Persephone is more *lunar* (and consequently more of a woman's conception of a woman)? First of all, she is obviously associated with darkness and the night due to her connection with the Underworld. More importantly, when she "came of age" (was no longer a virgin maiden), she took on the nature of periodicity, alternately concealed and revealed, like the lunar phases. (It is true that Venus can disappear behind the sun and change from Morning to Evening Star, but the periodic cycles involved are not the same as the regular monthly female biological phases.)[209] When a girl matures, she aligns biologically with the Moon (begins menstruating), and she enters into the woman's reality of eternal change, waves, ups-and-downs. But if this maturation is such a natural transition, why does Demeter mourn and search for Persephone? One answer is suggested in a story from the Mohawks of North America:

"When a young woman finds herself come to a state of maturity, she retires to conceal herself ... and when her mother or any other female relative notices her absence, she will inform her female neighbors, and all will begin to search for the missing one. They are sometimes three or four days without finding her, all of which she passes in abstinence."[210]

This lunar-death-and-rebirth (concealing-revealing) image seems consistent with some psychological views of women's behavior. One recently published, popular psychology book, *Men Are From Mars, Women Are From Venus,* has made the astute observation that women do not "think themselves out of an emotional upset," as men do. The author describes how women live emotionally in waves. When they are on a down-swing, he calls this "going down into their well." (We recall that when Demeter was mourning for Persephone at Eleusis, she sat next to the well; and the Demeter cult worshippers descended into crevices, caves, and the sites of underground springs in order to perform their ceremonies.) The author goes on to tell how other people (especially men) try to offer solutions to the woman who is "on her way down," but she will not respond. She must "hit bottom" first, before she can begin swinging up again. In other words, something old must die first; the woman must live out that death and rebirth emotionally by descending to the darkest depths, then re-emerging from her well, released and renewed. Despite the use of the name

"Venus" in the title of this psychology book, Venus the myth and the astrological planet has nothing whatsoever to do with this all-important aspect of female consciousness.

There is further evidence in Persephone's myth for the lunar connection. When she was abducted, her cries were heard by Hekate, who was in her cave at the time and invisible. Hekate appeared, torch in hand, nine days later, prepared to help Demeter search. During the interval Demeter had been searching for her daughter, crying at the crossroads. The nine days when Hekate was invisible and Demeter was wandering are the period of the dying moon, the last third of the month, and the time of the crossroad ritual. As we shall see in a later chapter, the Eleusinian celebration of the Thesmophoria was a period of abstinence for women that lasted nine days. Nine is one of the numbers usually associated with lunar cycles, never with solar ones, and it is of course the number of months in a pregnancy.[211] The three-part Goddess was a symbol of the different phases of womanhood or the Moon. Demeter is the Full or mature Moon; Hekate is the old crone, representative of the old, dying Moon; Persephone the new, waxing Moon. The New Moon could not be reborn without the old Moon dying. At least one ancient writer, Epicharmos, identified Persephone with the moon.

In historical Greece, the cult worshippers retained a lunar calendar, and in Hesiod's writings the life of the peasantry was regulated by the moon. Even today, farmers will look to the Moon to determine proper times to plant, prune, slaughter livestock, and hunters likewise observe lunar phases to indicate animal behavior.[212] It is highly likely that the annual storing of the seed-corn was grafted on to a more primitive observance which was monthly, not annual. We must remember that there is great emphasis in the myth on Persephone's stay in the netherworld being foretold by the number of pomegranate seeds she ate, and that most versions say it was one or three seeds. The early Greeks recognized only three seasons, so with Homer's version saying she ate one seed, the interpretation has been that this meant one season (of three or four months). But there is some indication that this one seed referred to the menstrual cycle of the dying moon.[213] Thus we can make sense out of the other versions which tell of three seeds, one for each month she is gone.

There are numerous references in ancient Greek literature to the herbal magic associated with female goddesses. Persephone and Demeter were constantly portrayed as holding or wearing poppies and pomegranates. The pomegranate was apparently a symbol of blood and fertility, yet Pliny tells us that the seeds were eaten roasted to arrest menstruation, and the women who observed the Thesmophoria were required to abstain from pomegranates and sexual intercourse. Paradoxically, the pomegranate was also sacred to Aphrodite, who promoted sexuality for pleasure and procreation, but who was herself more of an antagonist of motherhood than a ruler of childbirth. Pausanius remarked that "I will say no more about the pomegranate, because the story connected with it is in the nature of a secret."214 Another mystery: what was this one about?

As usual, there is a double meaning involved here. There are two varieties of pomegranate: white and red. The white kind was used to check menstruation, while the red presumably stimulated the blood. In the hands of Demeter-Persephone, the fruit had value for fecundity; with Venus, it could represent birth control. This makes sense when we remember that the orientation of Venus is towards the *pleasure* of sexuality, without the responsibility consequences of motherhood and child-raising (which she does not rule). Persephone, on the other hand, represents the *power* of sexual-fertility union, and with her, the pomegranate means rich, full, and powerful potential, complete with the implied responsibility we must accept for "up-ing" the energy level.

This argument can be related to problems we are now facing on the planet — depletion of natural resources and over-population. If we do not discipline ourselves with respect to materialism (Taurus) and curbing our procreative urges (Venus), then Persephone-Demeter will step in to prune our population via famine, genocidal wars due to migration, or with economic collapse ... all of which are beginning to occur worldwide. Likewise, if we do not act responsibly as caretakers of the planet, but continue exploitation through greed and gain-loss consciousness, we should not be surprised if Nature does what she must to revitalize and balance (Libra) the planet through the cleansing mechanisms of flooding, earthquakes, etc.

Returning to our discussion of the symbols, some sources say that the pomegranate sprouted from the blood of the slain Dionysos, and that to dream of pomegranates portended wounds.[215] The symbolism of the fruit is tremendous, contained energy which, when it is released, must be mastered. This is why the pomegranate was often regarded as a symbol of violent death, which is what happens (literally or metaphorically) when the consciousness is not mature enough to handle the release of energy. We see this as the inherent warning in our use of any powerful form of energy, whether it be nuclear or pharmaceutical. Rather than being cautioned to use wisely and with discipline, we are presented with the "may be harzardous" surgeon general's warning on cigarettes, the similar warning to pregnant women on alcohol labels, and those posted near radioactive, "harzardous" materials. Such laws and labels order us to avoid out of fear; they do not invite us to develop consciousness and self-disciplined moderation.

PERSEPHONE IN THE CHARTS OF
NUCLEAR, SEISMIC, AND UFO EVENTS

Astrologers have found many indicators for earthquakes, which include aspects to the angles and to eclipse points, strong Uranus or Pluto. It appears that Persephone-Transpluto aspects may be added as another important element to be investigated in earthquake research. In a survey conducted by the author, twenty-one earthquake charts were chosen arbitrarily for study. Sixteen of these charts (over 75%) showed Persephone making a hard angle to the Ascendant, with only very tight orbs used. Of these sixteen, all but one chart showed Persephone also in hard angle to either the Sun or Moon. We list a few examples of well-known quake charts (only tight orbs are considered):

In the 1906 San Francisco earthquake chart, Persephone is square the Ascendant and sesquiquadrate the Moon. In the 1989 San Francisco quake chart, Persephone is quincunx the Ascendant and opposed the Moon. The 1988 Armenia earthquake chart shows Persephone quincunx the Ascendant (and Ceres), and square to the Moon. The Tangshan, China earthquake of 1976 happened with Persephone conjunct the Moon, semi-sextile the Ascendant, sesquiquadrate the MC.

The 1963 earthquake in Skopje occurred with Persephone conjunct the Ascendant and semi-square the Moon. The chart for the 1899 Alaska earthquake (8.3 Richter) shows Persephone quincunx the Ascendant and semi-square the Moon. The 1811 New Madrid earthquake chart has Persephone sesquiquadrate the Sun and 15 degrees (semi-semi-sextile) from the Descendant. (Research by Theodore Landscheidt showed a relation between Transpluto and earthquakes, showing aspects of 7.5 degrees and multiples thereof between Transpluto and other planets.) A few scientists have suggested that the alignment of planets every 170+ years may be what triggers earthquakes. This period corresponds, of course, to one-quarter of a Persephone cycle.

Much more research needs to be performed to confirm this connection of Persephone with earthquakes, but these preliminary observations show a significant influence is present. This makes sense in terms of the myth of her being the daughter of Mother Earth. As supposedly caring humans, we become distressed over the destructive aspect of natural disasters and we forget the practical wisdom of Nature's pruning methods. Plate tectonics regulate the atmosphere's content of heat-trapping carbon dioxide over geologic time. Viewed without sentimentality and with greater vision as to the long-term survival of the planet, earthquakes are part of a system that continually renews most of the planet's crust.[216]

Earthquakes are evidently not the only natural disasters whose event charts show a prominent emphasis on this new planet. Persephone also makes her mark in the charts of floods, hurricanes, and volcanic eruptions. We only mention a few examples, stating once again that this is a fruitful area for further research. The chart for the infamous Johnstown Flood of 1889 shows Persephone conjunct the Moon; Hurricane Andrew of 1992 hit the Florida coast with Persephone semisquare the Moon and crossing the Ascendant. The 1991 eruption of Mt. Pinatubo in the Philippines occurred with Persephone conjunct the Midheaven and square to Pluto.

What is particularly fascinating is the prominence of Persephone in the charts for Nuclear Testing. For the first U.S. underground nuclear test, she is square to the Ascendant and making close aspects with five planets and the Nodes. In the chart for the first announced Soviet underground nuclear test (1962), she lies at the far midpoint of a complex of six planets in Aquarius. The chart for the second Soviet above-ground nuclear test (1951), Persephone is opposed the Ascendant. She often makes aspects with Pluto and Ceres in such charts, which makes sense when we think of her linking quality.

One final area for further research is Persephone's emphasis in the chart of UFO-related events and people. One example of this is Carl Sagan, self-appointed guru of science and one of the best known and most vocal of the debunkers of both astrology and UFOs, who was born Nov. 9, 1934. The chart for this date shows that Persephone had just begun retrograde motion at 29 Cancer 41, conjunct Pluto at 26 Cancer and Ceres at 0 Leo, with all three in square to Uranus at 28 Aries 45. This square indicates there is a definite problem here with relating to the alternative consciousness that is potentially present in the experience of meeting the greater "Other."

There is a great wealth of potential research material for astrologers in the charts of UFO sightings, due to the fact that exact times have often been recorded. One of these events is especially interesting because it has been called the "birth of the flying saucer."[217] That is the famous sighting by Kenneth Arnold on June 24, 1947, which was the first of about 800 sighting reports from the *UFO Wave of 1947.*[218] Arnold's report to the press gave birth to the media expression "flying saucer," and initiated UFOs as a phenomenon of great public interest. In the chart drawn for this sighting, Persephone is exactly square the Ascendant. The chart ruler Pluto is exactly conjunct the Midheaven and sextiles Venus, who is in turn semi-square to Ceres.

One curious aspect of the UFO phenomenon is the appearance of *Crop Circles,* the circular flattened areas of wheat and other grain crops which have been observed throughout the world, but most extensively in England. These are striking because of their symmetry and archetypal-symbolic nature. Several explanations have been offered for their origin, and there is much evidence connecting these circles with UFO sightings. Crop Circles began suddenly to appear in great numbers in the 1980s, and there are several obvious reasons for associating them astrologically with Ceres-Demeter and Persephone.

They are, of course, circles (cycles) in the corn. Serious researchers call themselves cereologists; CERES is also the acronym for one of the largest British groups studying the phenomenon. The symbolic patterns themselves have been often interpreted as suggestive of the Great Earth Mother (see figure below).[219] Crop Circles are another invitation to cross the "borders" that we think exist between scientific explanation and mythic-symbolic interpretation.

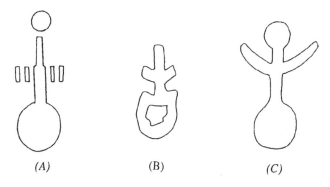

     (A)   (B)   *(C)*

(A) Crop circle pattern: 1990 England.
(B) Painting on a pebble: 9000 B.C. Spain.
(C) From a cave painting: 6000 B.C. Spain.

Science is only now re-discovering what the primitive mind has always understood: there is a "living" quality of the earth, even down in the "un-animated" rock. Scientists have recently discovered forms of microbial life deep inside the earth, causing suspicion that the planet has a *hidden biosphere* extending miles down. These new microbes, called hyperthermophiles, have been found in hot springs, active volcanic craters, and oil reservoirs, areas of extreme heat and pressure usually identified astrologically as Plutonian. Scientists say that the discovery of a deep biosphere would rewrite textbooks and shed new light on the origin of life itself.[220] This discovery, along with new findings about alien life forms and UFO contacts, is perfectly envisioned in the Persephonic metaphor.

Chapter 16

DISCOVERING A SYMBOL-GLYPH FOR PERSEPHONE[221]

A great majority of the transformations presently manifesting in the world are a result of a violent upheaval in our attitudes about the definition of "masculine" and "feminine." We have for too long amplified and exalted the so-called "masculine" principle of mind-over-matter. In order to achieve a workable balance, the so-called "feminine" principle has likewise been exaggerated, albeit to an opposite extreme.

The physical earth is itself near to a splitting point due to the stress exerted by the distance to which these dualities have become separated. From this ripping apart will come the realization of the possibility of a more androgynous and gender-unified concept of the Reality which we have presupposed was composed of opposite and separate dualities. This kind of thinking suggests the appropriateness of a mythological figure who embodies the struggle and celebration of both separation and reunion, of duality and unity. As we come to re-define what is masculine and what is feminine, we begin to discover the planet Persephone. Masculine mind and feminine feeling are both active in the birthing of this new planet. We shatter our traditional associations when we begin to think of BIRTH as *belonging to both the masculine and feminine realms.*

A woman knows she is going to give birth long before the emergence and externalization of the child ... this is called pregnancy or gestation. Likewise, humanity is now pregnant with the awareness of a new "arrival" in the family of planets. Mother Earth is in pain, but it is creative pain, the groaning and writhing of a woman in labor. Astrologers can assist as midwives to this emerging awareness by making birth "preparations" such as placing this planet in the charts of clients. Such actions help spread the consciousness of her presence, so that people can begin to identify her meaning within themselves.

Practically speaking, if we are to place Persephone in birthcharts, we require 1) a workable ephemeris to locate her position and 2) a symbol/glyph. As indicated previously, there exists a tentatively-accepted ephemeris which will suffice until greater accuracy is developed. Where there is yet no general consensus is on the matter of a symbol, except for the generic TP (Trans-Pluto) which is used for convenience in computer-produced charts. This TP recalls the early use of PL for Pluto (still a standard used by astronomers). Astrological planetary glyphs, like all symbols, are more than convenient abbreviations for chart-reading, however. They are projections of archetypal forces from the collective unconscious. They are not arbitrary and should not be treated in an offhanded manner. Each planetary symbol that astrologers use has emerged into common usage from the universal mind. We can assume that the same metaphysical processes that produced symbols for the known planets will also evoke a glyph for Persephone. The following is a discussion of a proposed symbol which succeeds, we feel, at expressing all that this new planet stands for.

Expectant parents usually choose a name for their child before it is born, and astrologers have similarly anticipated the new planet's name, curiously imitating most parents by arguing about the gender (Persephone if it's a girl, Bacchus-Dionysos if it's a boy). A new planet is like a child who must build its identity. The astrological identity of a planet is summarized and encapsulated in the picture of its symbol. And just like children of one family, the planets of our solar system share certain "hereditary" features among themselves. All the known planets have symbols which are comprised of one or more of three basic forms: the circle, the crescent (or half-circle), and the cross.[222] We can assume that Persephone's symbol will follow this tradition.

One symbol used by some astrologers for Transpluto is a circle containing a crescent or "equator" and crowned by an arrow (a variation on the cross). One astrologer has explained that this "winking Mars" indicates a directional pointing outside the known solar-lunar (circle-crescent) system. As such it is similar to a symbol once used for Uranus which represented a planet outside the then-known solar system.[223]

"Winking Mars" symbol
for Transpluto

Astronomers' symbol
for Uranus

The above symbol for TP-Transpluto gives off such a Martian air that it seems more appropriate for a male mythological figure. Although Persephone is certainly a strong and assertive personality (one would have to be to pair up as consort to Pluto), she requires a symbol carrying the power of the feminine nature. The sign she rules is air and therefore "male" or positive, but Libra is probably the softest, most receptive of the "masculine" air-fire signs. Similarly, Pluto-ruled Scorpio, though "feminine" water, is quite often mistaken as a fire (i.e., positive) sign.

Persephone, by way of her mythology, functions in two worlds, that of her mother Ceres and that of her husband Pluto. Because of this duality, we can expect to find in her glyph some duplication of one of the three basic forms underlying most planetary glyphs. The most feminine of these is the crescent. We can also expect the cross to be present, due to Persephone's strong connection with Earth. There will likely be a suggestion of similarity with Pluto's glyph because of the partnership-mirroring with him. Explanations were made earlier concerning the importance of Spica, the corn or wheat bud, in the concept of Persephone, and this too will be indicated in her glyph.

Persephone dies and is reborn in a constant cycling. She passes through stages from innocent maiden, to wife, to mother. Persephone is the seed carrier, reflecting the budding energy within that transports the seed of renewal and rejuvenation of life. Persephone's myth is a story of the maturation of the feminine, the growth of the bud-seed. All these concepts imply a glyph that is not static like a photograph, but a moving image (more like a film), one that develops and grows. Persephone travels between inner and outer realms, bridging the realms of her mother and her partner; she is ever on the move. Her symbol should therefore express the element of movement, migration, mutation. She is the mother of Dionysos, child of the dance. The emergence of her symbol is itself a dance.

We see the conception of her symbol beginning with the union of two primary forms, the circle (spirit) and the cross (matter), which merge and then split to form a new creation. This is not unlike the human process of the egg (circle) being pierced by the sperm (arrow-cross) and the first division into cells.

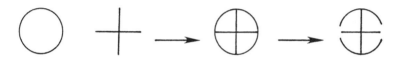

The circle has transformed into two crescents or arcs, and the cross has likewise split into two parts. There are now two mirror-image forms, a crescent below a T-square and a crescent arcing above a T-square. The shared horizontal line is the border between that which is above ground and that which is below.

The two halves of the crescent below close upward to form an oval-shaped bud or seed lying under the horizontal line of "ground level."

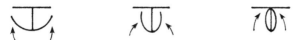

The two halves of the crescent above swing upward to form leaves opening from a stem above the ground.

The budding seed form rises up the stem and emerges as a bud between two leaves, just like an actual plant in its youngest stages of development. The symbol we have created resembles the corn stalk, the plant sacred to Persephone and Demeter.[224]

The bud is really two inner crescents facing toward each other, and the leaves are two crescents facing away from each other. Here we have the symbolism of the message to turn and look within, yet simultaneously be receptive to that which is "without." The bud, like many real seeds in nature, is not circular. This is symbolic of the non-circular (highly elliptical) orbit which has been determined for the planet Persephone.

Persephone returns to her mother every year in the spring and Demeter rejoices by making the earth green again. Thus do we return to unity with the earth. With the coming of autumn, the entire development of this symbol (as outlined above) is performed in reverse (like re-winding a film). The leaves wilt and drop to the ground in a downward direction, the seed goes back underground, descends to the lower world so that Persephone can reunite with Pluto. The never-ending cycle of death and rebirth continues.

There is another way to develop the Persephone symbol. Starting with the sign she rules, we can observe another "moving picture" of the symbol of Libra transformed into the symbol of Persephone:[225]

This symbol we have formulated for Persephone resembles the one for her partner. The "feeling" evoked by this symbol is a kind of "happy

Pluto," demonstrating the joy and freedom of life bursting up and outward. One of the names the ancient Romans gave to their Proserpina was "libera," meaning freedom.

The symbols for both Persephone and Pluto resemble the human figure. In Pluto's glyph, the "arms" are receptive and accepting ... not a typical description we associate with masculine gods. Pluto and his partner both mirror and reverse the traditional gender definitions. Persephone's "arms" are more assertive in expression. They resemble the pose of ancient "Orant" figures (see figure below); these are said to be the oldest and most natural posture for prayer.

The symbols for Pluto and Persephone are potentially interchangeable,

*Examples of the **orant** posture, said to be the oldest and most natural posture for prayer. Left: terra cotta figurine with pomegranate crown (Crete, ca. 1150 B.C.). Right: 5th c. A.D. sarcophagus figure.*

Illustration from photograph of African visionary dancer returning from trance state.

171

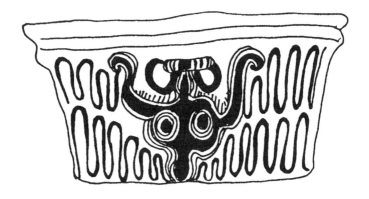

Pottery (Crete, 14th c. B.C.)

Mycenaean pottery designs (14th c. B.C.)

Late Minoan pottery designs.

demonstrating how we can mingle, exchange gender-identity, and transmute within our relationships. If Pluto's "arms" are moved downward until they cross the "cross of matter," and then rise upward, they "squeeze" the circle into a bud.

The keys to developing this symbol for Persephone are found in the initial, violent (earth-shattering) birth, the breaking up and reforming, and the movement. What appears to be violent and destructive is actually fuel for a new creation. This destruction and waste is what we have formerly associated with Pluto: death and compost. Alone (without Persephone), Pluto's "death" is frightening. There's no mothering aspect, no hope, no returning home. It's no wonder we have given to Pluto/Scorpio the rulership of death, possessiveness, attachment, jealousy. These are all variations on the fear of separation, loss, being alone.

By himself, Pluto is only half of the picture. Together, Pluto and Persephone are partners in death, united for the purpose of transformation. Pluto abducts us to the world of death, exposing us to deep fears and the experience of separation. Persephone reminds us that within the Universe nothing is ever lost, that the separation is only one part of an eternal cycling of together-apart-together again. She shows us another, more hopeful way of experiencing life: as wholeness.

174

Persephone's symbol, dubbed the "cornstalk," is rich in material for astrological and mythological interpretation.

Chapter 17

THE CELEBRATION OF ANCIENT MYSTERIES:
CALENDARS, PIGS,
AND ALTERED CONSCIOUSNESS

Mystery religions have played a significant role in the spiritual life of humanity. Unlike established state religions, they have expressed a concern for *who the individual is* in the context of a cosmic perspective, and they have offered redemption through a secret knowing *(gnosis)* of God. In ancient Greece the mysteries grew up alongside the myths, and in general, they occurred more usually in relation to the cult of women deities.[226] It so happens that one such mystery religion, called the Eleusinian Mysteries, were almost completely based upon the Persephone-Demeter myth, and were said, according to the popular traditions of Greece, to have originated during the 15th century B.C.[227] Homer's *Hymn to Demeter* explains the origin and significance of the Eleusinian Mysteries while relating the story itself, which we shall pick up where we last left off.

Persephone has been abducted by Pluto, and Demeter begins to wander in search for her. Ultimately, she arrives in Eleusis, where she goes to work disguised as a nurse for the child Demophon in the house of Celeus, living apart from the world of the gods. Among the humans she seeks the comfort of "substitution" by nursing a child whom she wants to make immortal, but her power is doubted by the child's mother. Abandoning her disguise and revealing her true identity, she demands the people of Eleusis build her a temple where she retires. Her mourning has now turned to rage against Zeus, and she withdraws the fertility of the earth in order to prevent people from sacrificing to the gods.

The myth goes on to tell how, at Celeus' house, Demeter first broke her mourning fast by drinking *kykeon,* a combination of barley meal, water, and mint. She rewarded the good people of Eleusis for building her temple by teaching them her mysteries, which they passed on to the following generations. Although there were other shrines built to Demeter and Persephone in Greece, most notably in crevices and caves, the temple at Eleusis remained the one place where the Mysteries were celebrated. All this is told in the Homeric Hymn, which concludes with praise for the virtues that come to those who participated in the Mysteries. "Happy are men who have seen these things." So important were these Mysteries to the ancient Greeks that, as one scholar wrote, "a thousand years later, when Christianity put an end to the mysteries of Eleusis, Greek life itself seemed to have sunk into the grave with them."[228]

The astrological timing of this latter event is worth discussing. The Mysteries were continuously celebrated until their suppression by the Christian Emperor Theodosius in 389 A.D. (Theodosius was very busy at this time outlawing every religion but Christianity.) The sanctuary at Eleusis was burned by the Goths at the end of the fourth century.[229] Earlier in this book we made mention of the importance of sign ingress of the outer planets for marking major cultural change, and this period of the late 4th century was no exception. All in one year, 389 A.D., Pluto entered Taurus, Neptune entered Aries, Uranus entered Sagittarius, and Saturn entered Virgo. By the end of the 4th century, all four outer planets, plus Jupiter, had entered earth signs, and Pluto was situated at the powerful crosspoint of 15 degrees Taurus. Persephone was throughout this period in the "Christian" sign of Pisces and rather swamped (or muddied) by the heavy authority-materialism of this outer planet emphasis on Taurus-Capricorn.

Returning to our discussion of the Mysteries themselves, we recall that the images of Demeter and Persephone are powerful in their myth, and the Mysteries attest to this. Demeter was not merely the weeping, grieving mother searching for her lost one, but majestic and highly principled. She could be wrathful, unyielding, and vindictive. And perhaps of most immediate concern to the common people, she could assure the good harvest. (We shall see that this "good harvest" was a consequence of proper sowing, which is really what Demeter ruled.)

Demeter has been occasionally associated in modern interpretations with the classic image of the *Mater Dolorosa*, but the combination of grief and rage is not typical of this prototype (such as the Virgin Mary, who is exclusively grieving and never angry). This has been the common portrayal of the feminine principle, rage being split off and projected onto witch-like figures. But clearly, Demeter retains both emotions. She is not a fragmented feminine figure.[230]

Persephone is equally complex, for in one aspect she is the innocent virgin Kore who is abducted and mourns separation from her mother and the earth, but finally she matures and appears to become powerful during her sojourn in the underworld as the Queen of Hades, and is called "the one who slays all things" and the "dreaded Persephone."[231] She journeys to the depths and returns radically changed, identified with the soul which survives death.[232]

A third significant goddess connected with the Eleusinian mysteries is Hekate. She is often overlooked in mythology, but is present throughout the Persephone myth, first accompanying Demeter on her search, later becoming the companion of Persephone. In Greece the Triple form of divinity was exclusively confined to the female goddesses.[233]

The Mysteries themselves were divided, like the Double Mother-Daughter Goddess, into two parts, the Greater and Lesser Mysteries. The Lesser Mysteries were celebrated in the Spring during the month of flowers, the *Anthesterion*.[234] The Greater Mysteries were held in the Fall, and included the festival of *Thesmophoria* ,which was only for women. This celebration included re-enactments of Persephone's descent by the participants descending into crevices in the earth and then returning. Pigs were used in this ritual because they had been swallowed by the earth when Persephone was originally abducted. The festival got its name from *Demeter Thesmophoros* (Demeter the Law-Carrier or Law-Giver) which reflected not only on the Persephone-Libra rulership of justice and law, but also on how the peaceful nature of agriculture had brought about a moral social structure to the ancient world. It was understood by the Greeks that the arrival of Demeter in Greece had brought agriculture, settled life, marriage, and the beginnings of civilized law.[235]

The Greater Mysteries were held once a year, but every fourth year with particular splendor in what was known as the *penteteris*.[236] A holy truce was proclaimed throughout Greece and the ancient world, and all warring ceased so that participants in the Festival could travel to and from Eleusis in peace. The truce lasted 55 days, from the fifteenth day of the month of *Metageitnion,* through the entire month of *Boedromion,* to the tenth day of *Pyanepsion*.[237] The month of Boedromion (corresponding to late September-early October, i.e., the period when the Sun, to moderns, is in Libra) was considered to be the real month of the Mysteries, and the period before and after (Solar Virgo and into Scorpio) was allowed for preparation, consummation of the rites, and travel across borders to return home. We note that these Zodiacal correspondences are completely in accord with our previous discussion of the side-by-side sign rulership of Ceres, Persephone, and Pluto.

Careful archaeology and research has made it possible to understand at least some of the rituals and cultic practices of the Mysteries. The Greater Mysteries lasted nine days.[238] The first day the people assembled; the second day began with purification in the sea. Each participant carried a pig which was also cleansed, later to be sacrificed and eaten. The third day was given to Prayers. The fourth day was rites for latecomers. The fifth day was celebrated with the grand procession from Athens to Eleusis. The sixth and most significant day began with rest, fasting, and purification in preparation for the evening. The fast was broken by drinking *kykeon.* Then the initiates apparently went through certain experiences which left them filled with awe and confusion, but also overflowing with bliss.

This same evening was devoted to a sacred pageant dealing with the abduction of Persephone, the wandering of Demeter, and the reunion of the two Goddesses. But also (and this is the part of the Mysteries that have been kept a secret for 2000 years), the initiates were shown holy things and were witnesses to events which were evidently not a dramatic play but a divine presence that realized the myth.[239] The seventh day was spent resting up for a second level of initiation. The eighth day was given to rites for the dead. The ninth day marked the beginning of the return of the initiates to Athens or their homes.

The Eleusinian Mysteries were unique in that they celebrated the relationship between mother and daughter. Religions have often celebrated the mother-son; Carl Jung noticed this and wrote, "the Demeter-Kore myth is far too feminine to have been the result of an anima-projection."[240] Jung also wrote, "Demeter and Kore, mother and daughter, extend the feminine consciousness both upwards and downwards ... We could say that every mother contains her daughter in herself and every daughter her mother, and that every woman extends backwards into her mother and forwards into her daughter ... The conscious experience of these ties produces the feeling that her life is spread out over generations, the first step towards the immediate experience and conviction of being outside time, which brings with it a feeling of immortality."[241]

The Eleusinian Mysteries were also precisely that: *mysteries;* although scholars have figured out certain rites, we have little proof of the actual content and substance of the Mysteries. Initiates were sworn to secrecy, and they kept their vows well. We do know, however, that the cult of Eleusis appeared to satisfy the most sincere yearnings of the human heart. Not only Homer, but all the great philosophers seem to indicate this. Sophokles wrote, "Thrice happy are those of mortals, who having seen those rites, depart for Hades; for to them alone is it granted to have true life there." Pindar said, "Happy is he who, having seen these rites goes below the hollow earth; for he knows the end of life and he knows its god-sent beginning." And Cicero wrote that Athens had given nothing to the world more excellent or divine than the Eleusinian Mysteries.[242]

THE SACRED CALENDAR

The dating of the Lesser and Greater Mysteries, and their relationship to the Attic Calendar, is worth discussing for our interpretation of the Persephone myth and its relevance to astrological timing.[243] It has been widely thought that Persephone is associated with the return of spring, the seed that begins new growth, and that Demeter-Ceres represents the grain harvest. In terms of symbology translated to the northern European and North American climate, this is an acceptable interpretation. But it is a mistake to assume that the Greater Eleusinian Mysteries celebrated a fall harvest. The climate throughout the ancient

Late Bronze Age plaques of fertility goddesses holding wheat stalks.

Greek world was such that the agricultural season was very much reversed, as it is in some southern parts of the United States (like Arizona, where seeds are sown in the late fall and the harvest in reaped in the spring), or in the case of "winter wheat," which is planted before snowfall and harvested in early summer. There was, in fact, no harvest festival celebrated strictly in honor of Demeter in ancient Greece, but she figured large in the festivals of sowing-time.

There are many theories as to how the Mysteries were connected to the agricultural year and the calendar. What has continued to cause confusion is the important characteristic of duplication — of divine figures, of the mythic and ritual events, and the motifs. What we must do is be open to several possible explanations simultaneously, for this seems to be the way to deal with and understand the nature of Persephone and Libra. Being unable to choose only one viewpoint, formula, or interpretation (because they all seem to contain some portion of the truth), we choose "both" or "some of each." One example of the inherent contradiction in the myth which makes it difficult to correlate with the calendar is this: The myth fixes the time of Persephone's abduction as the spring, the time of blossoming of flowers, as well as for the symbolic freshness of the young maiden. Yet her departure is mourned in the fall, when Ceres makes the earth stop growing. The earth turns green again in the spring, because Ceres is joyful at her daughter's return, yet the spring is the time of her abduction. In the agricultural-symbolic sense, Persephone Kore the seed is raped and carried off to "death" when the seed is harvested and stored in the ground, which should be a time for rejoicing that "the crop is good." And mixed into all this repetition is the question of whether the sowing and harvesting are performed in the spring or fall.

According to one view, the Mysteries could be the occasion of Persephone's return, since Kore was the Grain Maiden who was reaped (raped) at harvest time and put down into the underworld of Pluto, the storage pit for the seed grain. After this sojourn, Persephone becomes fertile and is ready to be planted and to produce more grain. This interpretation says that the Mysteries are celebrated when the grain is taken from the granary (brought up out of the ground); this is supported by the fact that the ear of wheat was one of the holy objects shown to Eleusinian initiates. But being underground does not make a seed more fertile; in fact it can rot if the farmer isn't careful. The metaphor of rebirth is derived from the seed's apparent death and rebirth, but not from the storage per se.

Another way to look at this is to consider the practical farmer's view. The season of sowing is a time of great anxiety. The stored seed might have gone bad over the summer and might not germinate. The farmer must decide which crops to plant and in what amount, and how much grain must be saved to be eaten over the winter. The decisions made at sowing time may determine what happens for a whole year, and even prudence and the wisdom of experience cannot stand up to the elements and the whims of nature. It is not until early spring that the farmer knows that his crops will be a success. From this viewpoint, it is possible to conclude that humans might feel a greater need for ritual and the blessing of the Goddess at the more insecure time of sowing, more so than at harvest time, when they would nonetheless be offering praise to the Goddess. At harvest time there is reason to be grateful to Demeter, but the motivation of insecurity and anxiety which supposedly urges the need for ritual is absent. Astrologers are familiar with this behavior: their clients consult them in times of doubt, not when things are going well.

The story of Persephone's ascent and descent can be applied to the agricultural year; it just depends on which assumption is made first. We can identify Persephone as the grain, which dies or goes underground at autumn sowing time and is reborn in the early spring. Or we can identify the time of Persephone's "death" with the reaping of the grain in late spring-early summer (in ancient Greece) and its deposition underground in the granary. In this latter situation, her resurrection would then be in the fall, when the seed was sown and became fertile again. Neither view can manage to embrace the whole myth, but perhaps this is another proof that myths are not intended to have exact and precise application.

Myth has been described as something which grows out of ritual. It has also been said that ritual and myth express the same idea; that ritual is not intended to have a practical effect, but is simply symbolic of the divine level. However, the emphasis of the ancient festivals on sowing rather than harvest seems to imply that, the more critical the situation,

Goddess giving birth, guarded by scorpions (Ur, ca. 2400 B.C.)

Hellenistic figure of *Dea impudica* (the "shameless goddess"),
exposing herself and riding a pig.

the more vital is the ritual. Agrarian rituals were performed with the purpose of helping to influence natural events. The festivals of ancient Greece seemed to emphasize the autumn sowing and first "greening" of the fields in the spring. Likewise, the Homeric myth put greater stress on the Kore's abduction and her yearly return. Looking at this syncretism, it would appear that we could, with some confidence, interpret Kore's descent as in the fall (Solar Libra-Scorpio), and her ascent in early spring (Aries-Taurus).

THE SACRED PIG

An important feature to all the festivals of the Eleusinian Mysteries is the ceremonial role of the pig. Part of the ritual of the Thesmophoria, the festival dedicated to Demeter-Persephone and celebrated exclusively by women, was the sacrifice of pigs, which were tossed into crevices or pits in the earth. Some sources give an earlier time for sacrifice and burial; these carcasses were allowed to rot, were dug up three months later, placed on altars, and mixed with seeds that were used for sowing of the new crops in September-October.[244] (Here again, we have the "one-third" of a year symbolism. The grain was harvested in June, concurrent with the sacrifice of pigs; the seed and the sacrificed pigs were both stored underground for 3-to-4 months, then rejoined at sowing time. Since the pig was representative of Demeter, and the seed was Kore, the reunion of mother-daughter was complete.)

In related rituals, the pigs were bathed and purified, pig's blood was also used for anointing in purification rites, and grain cakes were made in the shape of pigs. So important was the role of the pig in the Eleusinian Mysteries, that when Eleusis was allowed to issue coinage in the 4th c. B.C., the pig was chosen to be the symbol on the coins. One of the names for Persephone was *Pherrephata*, "killer of suckling pigs."[245] Not surprisingly, these animals also figured prominently in the myth of Persephone. When Persephone was abducted, a swineherd was tending his pigs in the field. The earth opened up, the pigs fell into the chasm as Pluto and Persephone vanished below, and earth closed over them. The swineherd Euboleus appeared in other, later myths as Persephone's earth-time husband, an adaptation which probably reflects the patriarchal tendency to assign husbands to independent female goddesses.

There are many reasons why the pig makes sense as the animal most sacred to Demeter and Persephone. The pig had long been worshipped in primitive, matriarchal, agricultural societies. Before the invention of the plow, the pig's habit of rooting about in the soil made its identification with agriculture very natural. Anyone who has raised pigs can cite all the benefits: if enclosed in a field, they will remove rocks and weeds, and fertilize the area, making it ready for crops. The spreading of pigs' blood provides further nutrients to the soil. Pigs are fast-growing and provide a multiplicity of offspring. They appear throughout the world as one of the common, combined symbols of fertility, death and rebirth. Even the Jewish and Muslim taboos on eating pig flesh seem to have arisen from an ambivalent attitude about the divine nature of the animal. Modern rationalization would have us believe that pig meat was avoided in Biblical times because of its "uncleanliness." But was it forbidden because of its sanctity? There is evidence that Pig was not eaten because it was divine.[246] But since its divinity was "of the Goddess," we can see how such patently patriarchal religions as Judaism and Islam would discriminate against the pig. The Jewish and Muslim aversion to the pig was a "deliberate rejection of the practices of 'heathens,' for whom the pig stood for fertility and wealth and who thus chose it as a sacrificial animal and pork as a special food."[247]

This is confirmed by the observation that the disappearance of the pig as a domestic animal in ancient Mesopotamia was contemporaneous with the lowering of the social status of women. The first written evidence of women's subjugation in farming communities dates from about 1100 B.C. in Mesopotamia.[248] One anthropologist has explained how general female subordination coincided with the development of *plow* agriculture, and this theory brings together several elements that we have mentioned: the rooting mechanism of the pig, the late assignment of a swineherd earth-husband to Persephone, and the fact that the Homeric Greeks considered Demeter, the goddess of agriculture, as the one who brought civilization, law, and marriage into society:

"Culture often complements nature's laws, and the low divorce rates seen in preindustrial European societies were also due to an inescapable ecological reality: farming couples needed each other to survive. A woman living on a farm depended on her husband to move the rocks, ... and plow the land. Her husband needed her to sow, weed, pick, prepare, and store the [crops]. Together they worked the land. More important, whoever elected to leave the marriage left empty-handed. Neither spouse could dig up half the wheat and relocate. Farming women and men were tied to the soil, to each other, and to an elaborate network of stationary kin."[249] But prior to the plow, farming was accomplished with a hoe. "In cultures where people garden with a hoe, women do the bulk of the cultivating; in many of these societies women are relatively powerful as well. But with the introduction of the plow—which required more strength—much of the essential farm labor became men's work. Moreover, women lost their ancient honored roles as independent gatherers ... and soon after the plow became crucial to production, a sexual double standard emerged among farming folk. Women were judged inferior to men."[250]

It is our contention that the early version of the Persephone myth and the Eleusinian Mysteries give a picture of pre-patriarchal societal conditions. They relate how, prior to the plow, the *pig* assisted women in working the soil, thus providing enough labor to allow women to remain independent and powerful in their own right. With the development of plow agriculture, the pig was "replaced" by the husband, and the entire panorama of Western sexual prejudice was set in motion: men had more important economic roles as farmers, couples were obliged to remain together in their mutual dependence of "marriage to the soil," and society gave even more importance to men because it needed political leaders, diplomats, warriors, and traders to defend and manage that soil property. It is not a new idea that patriarchy, capitalism, and male monotheism are inter-related and responsible for the exploitation of nature and women, but we get an interesting angle on the subject with the Persephone myth.

With the development of plow agriculture and the shift to patriarchy, the female was not only divested of her original role and connection to nature, she was instructed to transfer her obedience, respect, and worship of natural forces (like the pig) to the new head of household, her husband, as well as his male god and related socio-economic structures. We can see how pigmeat became taboo; the pig was obviously too closely associated with the power of the independent

Iambe-Baubo figures from Asia Minor, 5th c. B.C.

feminine for the "taste" of the patriarchy. This pun brings us to the
touchy subject of who exactly defines *good taste* versus *poor taste,*
how the myth of Persephone has been sanitized and divested of its
primitive "crudity" by modern morality, and what archaeological
evidence shows about the supposedly depraved and obscene nature of
the Eleusinian Mysteries.

BAUBO AND THE VIOLENCE OF WELL-BORN LADIES[251]

At the end of the nineteenth century, in the remains of a temple of
Demeter and Persephone, archaeologists discovered a strange group of
terracotta figurines (see example, next page). Each one of them had a
disproportionately large head sitting directly (without a torso) on top of

legs, and below the mouth was a representation of a woman's genitals. These objects were soon recognized as being related to Baubo, the woman who nursed Demeter. Lexicographers understand the word *baubo* to be related to the word that means "cavity" (*koilia,* a term that indicates both the belly that digests and the belly that conceives). The masculine form of this same noun, *baubon,* refers to a leather phallos, i.e., a false phallus or dildo, an object that was praised in Aristophanes' *Lysistrata.* The mythological figure Baubo was variously identified by different ancient sources as queen, nurse, slave, and "demon of the night."[252]

Victorian scholars (and their modern but still straight-laced cousins) were alternately disturbed and titillated by the accumulating evidence about Baubo and the lewd behavior of women involved in the rituals of the Eleusinian Mysteries. Statues were discovered that depicted Baubo in various improper positions or riding a pig. It was clear from the ancient texts that the Mystery initiates followed a sequence of ritual: they fasted, they drank the *kykeon,* and they were shown sacred objects, which they handled in some way, returning the objects to a basket, and then to the *kiste* (sacred chest or coffin). Various ancient authors had openly identified some of these objects as grain seeds, pomegranates, bread cakes in the shape of pigs, all of which were clearly associated with the Persephone myth. But they also made less definite allusions to another type of "handled" object which scholars interpreted to be the following possibilities.

One suggestion was that the object was a phallus which the initiate placed on his breast, uniting himself with the Goddess, but many scholars refused to admit the phallus as part of the Eleusinian cult. Another suggestion was that the object was not a phallus but the female pudenda, the *kteis.* Going a step further, one scholar maintained that the initiate actually came into symbolic union with the Goddess by manipulating his own genital organ in the kteis. Another projected that a mystic union was effected by manipulation of two objects in the basket, a phallus and a kteis. Other scholars decided on a "you can look but don't touch" concept, that the objects were simply shown to the initiate.[253] The scholarly debate over this issue got so literally "hot and heavy" that it elicited a conservative response by one of the most respected scholars in the field, who reminded others that "chastity was a prerequisite for initiation" into the Mysteries at Eleusis, that "the handling of the phallus and of the kteis is an unclean act," and that, since children were included among the initiates, "to expose these children to the act would have been criminal."[254]

Aside from this scholarly debate, and what it reveals about modern morality, what is interesting for us in the figure of Baubo is that she tells us something about what made Demeter laugh and cease her mourning. The Homeric Hymn gives credit for the release of the Goddess from mourning to the words of a woman named Iambe (elsewhere called Baubo) who transformed the mood of the divine Mother. Let us look at the story in more detail. In the course of her wanderings, Demeter had disguised herself as an old woman and arrived at Eleusis. At first she refused the food and drink offered to her, and likewise abstained from any word or gesture. Enter the character Baubo, who is depicted as a crude and rustic old woman of the backwoods. She informs Demeter that she is an uneducated peasant who talks too much, and that she has no gifts of honor to offer. After exhausting the arguments of rhetorical persuasion, Iambe-Baubo says, "On the other hand, if you are willing to loosen the bond of your mourning, I can set you free."

Other sources go on to describe how Iambe-Baubo began to use licentious language, insults, and obscene jokes to cheer the Goddess, and then suddenly pulled up her dress to expose her genitals. Demeter was delighted and "smiled in her heart." Thus, "the obscenity of a lewd act was able to attain what Baubo's [previous] modest behavior had long failed to accomplish." Some of the texts relate the details: "Baubo exposes to sight the objects shaped on her natural parts, agitating them with her hand. The shapes resembled a little child, and she pats it and gently manipulates it." After having sat without smiling for such a long time, Demeter finally laughs, relaxes, and breaks her fast with the drink made of barley, water, and pennyroyal mint, which is offered to her by Iambe-Baubo. The two form a lifelong bond, which is attested to in other myths.[255]

The various modern approaches to the Baubo affair have naturally given rise to considerable debate and controversy, even to a point where some scholars have banished the disgraceful story from Eleusis altogether. Others sought to provide Baubo with the euphemism of a "legitimate" spouse, calling them together a "shy and courteous" couple or burying them in "the heritage of a very brutal past." Some scholars have avoided the whole problem by reducing Baubo to the status of being just one of the "riddles" of Eleusis.[256] It has been pointed out that "this sort of archaeological reconstruction of ritual receives its principal legitimation from certain psychological assumptions and from a vision of 'primitive' religious mentalities,"[257] namely, that obscene or sexual representations and rites of agricultural fertility or mystical rebirth are defined as having an archaic origin. This "archaic origin" is, in turn, viewed as primitive and undeveloped, and makes possible the historic model of Progress, Civilization, and (we might add) "good taste." A simple look around the pornographic reality of modern "civilization" will expose this historical theory as both condescending and illusory.

But there is another way of explaining Baubo, which follows (instead of avoiding or altering) the obscenity track. Historians and philologists have discovered many links that existed among the Demeter-Persephone myth, Baubo, Eleusis, the Thesmophoria, and many other women's festivals. The Eleusinian celebrations included events attended only by women, such as processions where women hurled insults and howled obscenities or dirty jokes, and ceremonies that included the handling of sexual objects. Ancient sources described these events as the exchanging of indecent language, mockery, blasphemies, shamefulness, obscenities — and always among women. Men (and in some cases, even male dogs) were excluded from participating.

One socio-psychological interpretation of this behavior is that women in Greek times led restricted lives, and that, among themselves and away from the inhibiting presence of the opposite (and dominant) sex, they would naturally "let it all hang out." The comparison with the modern day example of the bride's shower or the bachelor's party is cited in support of this view, demonstrating that behavior becomes more daring and lewd under the influence of intoxicants, the absence of the opposite sex, and the occasion of the "last fling" of freedom before the impending ritual passage into restricted monogamy. One criticism of this analogy is that it is based on the same attitude that underlies the false portrayal of the "happy Negro slave" who is restricted yet obedient in the presence of masters, and finds release through gospel singing, tap dancing, voo-doo, and attending holy-roller church. It is the same attitude that assumes the subservience and inequality of women can be relieved and justified by a trip to the shopping mall or the hairdresser.

It is often difficult to extract the real nature of a myth from the layers of patriarchal revisionism. This can be seen in the contrasting views held by the Greeks of male and female roles. One modern interpretation says that Baubo as the personification of the vulva is the counterpart to Priapos, the incarnation of the phallus. But the Greeks assigned different language to acts, depending on the gender of the participant. When the male Priapos exposed his genitals, it was "rustic crudeness" or "grossness." If a woman acted in the same way, she committed a "shameful" act.

We might gain an understanding of this complex issue by reviewing again the details of the Persephone myth. Demeter is mourning, which Baubo succeeds in ending by mockery, ridicule, indecent exposure and language. When Demeter laughs, the end of her despair heralds the immediate return of the fertility of the soil, and by association, human reproduction. It makes perfect sense that the behavior which consoled Demeter would be repeated in the so-called obscene rituals celebrated in honor of Demeter and Persephone, and that the hope and aspirations for renewed fertility of the earth would be assured by the enactment of obscene activities that suggested sexual reproduction.

Demeter is further linked to Baubo because, disguised as a nurse, she took over the role of raising Demophon, whose original wet nurse had been Baubo. In different sources Baubo is identified as the nurse to the mother of Persephone. Another ancient writer has described this cross-identifying and "mirroring" process: "Demeter laughed to see Baubo exposing herself because she recognized herself in Baubo."[258] Here is the crux of the matter: we suffer and mourn, but when we *see ourselves in others* (or others in ourselves), we can let go of the grief and rage to which we've been attached, and go on with the etcetera of life.

We can surmise that the involvement of Baubo in the myth of Persephone may be important to the meaning of the astrological planet Persephone. The emergence of this planet into general consciousness is certainly reflected in current trends such as the growth of the feminist movement and the mass gatherings of women, the wide publication of pornography, the open display of obscenity (as in RAP music), and the hurling of insults that accompany the moral debates over the freedom of expression (including sexual expression), family values, and abortion.

Baubo has demonstrated one method for breaking through to new consciousness, the shocking effect of humor that comes from seeing "the naked truth" about ourselves. But there is another way to alter consciousness ....

THE ELEUSINIAN VISION QUEST:
HOW THE INITIATES GOT HIGH

All the ancient texts agree upon one fact: "something" was shown to the initiates participating in the Mysteries at Eleusis and they were sworn to secrecy about what they saw. The same words are used over and over to explain this: *No one could reveal what had happened there.* The secret of Eleusis was not only a secret of which it was forbidden to speak (*aporheton* in Greek), but also that which could not be spoken (*arrheton*). Persephone was called *arrhetos koura,* "the ineffable Maiden," the only deity to be so called.[259] Most scholars have assumed this meant that initiates were not *allowed* to relate the mystery, but what if the truth of the situation was that they simply were unable to explain it? What if the "something" they saw was not literal objects in a basket or a dramatic presentation of the myth, but a *vision?* One group of investigators (who included Albert Hofmann, the man who discovered L.S.D.) has come up with a unique and relevant explanation.[260]

These investigators started with the evidence. They noted first that all those who had written of the Mysteries, such as the poet Pindar and the tragedian Sophokles, testified to the overwhelming nature of what was seen by initiates, saying that the experience itself was incommunicable,

193

for there were no words adequate to describe it. The closest that Pindar could get was to say that he had seen the beginning and the end of life and had known that they were one, and that the division between earth and sky melted into a pillar of light. Secondly, the sources indicate that the initiates experienced physical symptoms accompanying their experience: fear and a trembling in the limbs, vertigo, nausea, and a cold sweat.

These seemed to the investigators to be symptomatic reactions, not to a drama or ceremony, but to a mystical vision. They reasoned that, since this vision could be offered to thousands of initiates each year on a regular schedule, it seemed obvious that some kind of hallucinogen must have been used to induce it. There were two other observations that supported their argument: a special potion (the kykeon) was drunk prior to the experience, and the historical fact that a notorious scandal had occurred in Athens when it was discovered that numerous aristocrats had been "celebrating the Mystery" privately in their homes during dinner parties.

The more the investigators looked, the more evidence they uncovered to support this theory. In the myth, the flower that Persephone had picked was the narcissus, which the Greeks themselves thought was so-named because of its narcotic properties. (Our own English word "narcotic" stems from the Greek *narkissos.*) The common theme of the abduction of maidens while gathering flowers was discussed by Plato; he noted that the usual companion of such seized maidens was named *Pharmaceia* (meaning "the use of drugs"). Considering this, Persephone's abduction could be seen as a drug-induced seizure. There is also the archaeo-artistic evidence of numerous sculptures, stele, and pottery depicting the Persephone figure crowned with opium seeds (see figure, page 194) or accompanied by the poppy flower.

There is also the related figure of Dionysos, whom we know to be the god of the vine and the wine. All the descriptions of this god tell of an inebriated state that surpasses what we know to be the common effects of alcohol ... it was called raving madness, and could not have been produced by wine alone, especially considering the fact that the Greeks drank their wine diluted. The myths and artistic representations of Dionysos always include female devotees gathering herbs and flowers, which other sources explain were added to the diluted wine. These maenads had as their emblem the *thyrsos,* a fennel stalk into which they stuffed their cuttings. This gathering was always performed in the early spring, just preceding and during the time of the Lesser Mysteries of Eleusis. In every ancient source, it is made very clear that these Lesser Mysteries were considered a required preliminary for the vision of the Greater Mystery experienced in the fall, at the time of the autumn sowing. The Lesser Mysteries were observed during Anthesterion, the Flower Month, which is the time in Greece when bulbs bloom (February).

Now we come to the potion which Demeter and the initiates drank, the *kykeon.* Its ingredients were water, barley, and mint, a seemingly harmless combination. The mint will first be discussed, then the barley. The plant which was used was pennyroyal, a mint with mild psychotropic properties. Aristophanes refers to *blechon* (pennyroyal) as found in an aphrodisiac; he also uses it as an obscene metaphor for woman's pubic hair. But pennyroyal is also an abortative. In ethnobotanical lore, mint was said to be Pluto's concubine, dismembered by Persephone. It was also said that Demeter showed her repugnance for an illicit union by grinding mint underfoot and condemning the woman to perpetual sterility. We are beginning to understand that a barley-and-mint concoction is not as innocuous as it first sounds. We now turn to the grain ingredient.

And here we must digress into a little history about the cultivation of grain. Grain was developed as a hybrid of a wild grass, and in the early stages of its cultivation, grain was likely to revert to its primitive form if not tended properly. This original grass, called *aira* in Greek (*lolium temulentum* is its botanical name), was susceptible to a fungus called *ergot* or "rust." Thus we have the symbolism of the reverted wildness of Dionysos and the fungal abductor Pluto, both potentials for endangering the seed grain (Persephone), the offspring of cultivated and civilizing Demeter. Ergot or rust (botanical name *claviceps purpurea)* is visible as a reddening corruption to which barley is highly susceptible; its visibility is further observed when the sclerotia fall to the ground and produce tiny purple mushrooms. An ancient Greek comedy fragment describes the crumbs of barley with which a drunken potion was created as "purples of barley."

The influence of this red rust or purple ergot is a powerful hallucinogen. The ancient Greeks were quite aware of the psychotropic properties of *aira*. They called it the "plant of frenzy." We can imagine what happened to the initiates after they drank their potion. Among other effects, ergot disturbs the inner ear (the sense of balance) and produces astonishing ventriloquistic effects. Ancient sources refer to one of its effects as seeing things that are not there. Pliny recorded that bread made from contaminated grain caused vertigo.

It appears from mythical sources that one of Dionysos' botanical transformations was some kind of mushroom and that the "winter bulb" sought out during the time of the Lesser Mysteries was just such a fungus. There is evidence that indicates that the "winter bulb" was the characteristic mushroom that grew from the ergot of the grain harvest. It was phallus-shaped and associated with Dionysos. The worship of pigs seems to further support this theory. Pigs are known for their ability to root out mushrooms with their snouts. They are used in France even today to search out and uproot the highly-valued truffle.

Between Dionysos and Demeter is the symbolic opposition between wild and cultivated plants. Fungi are seedless growths that defy cultivation and are therefore emblematic of wildness. Moreover, they feed upon dead and putrefying matter (a Plutonic element), and no doubt held a special relevance for the ritual of rebirth. The Greeks thought of the tomb and the whole underworld as covered with moldering growths that consumed the flesh of mortality.

Thus, parallel with the Dionysian gift of wine and mushroom was the grain of Demeter, which could be transformed by a wild fungus. The vision that resulted from the ingestion of ergot demonstrated the continuity of life and death. The inclusion of mint in the Eleusinian potion added more connotations of illicit sexuality, symbolic of the transition of Persephone into wifehood. Persephone's matrimony was a joining of the world of the Olympians (the sky gods) to the netherworlds. The sacred reconciliation gave humanity an intermediate role communicating with the chthonic powers for the fertility that would yield the continuing cycles of rebirth.

The Mysteries were double, just like the holy duo of goddesses, Demeter and Persephone. The emphasis of the Greater Mystery was on redemption from death by incorporating the underworld's putrefication so that it functioned as a source of fertility — as *manure instead of pollution*. The Lesser Mystery was more wild and involved the Dionysian theme of gathering intoxicating uncultivated flowers prior to the abduction of Persephone.

THE ASTROLOGY OF CEREMONIAL CHEMISTRY

Because there is a strong connection between Persephone and altered consciousness, we would expect to see the astrological planet Persephone as prominently involved in drug-related circumstances and the charts drawn for such events. Traditional astrology gives the rulership of drugs and intoxicants to Neptune, the ruler of mystical and altered-state-prone Pisces, so we might see a combination of these two factors. We offer a few examples from history to illustrate our point, and include all sorts of drugs, chemicals, and intoxicants:[261]

Persephone's ingresses seem to be a particularly important factor in the timing of drug events. The early 1840s were not only the time of Neptune's discovery but also the ingress of Persephone into a new sign, Gemini. The first *Opium* Wars were fought 1839-1842. *Hashish* was first used to treat mental patients in 1841. *Cocaine* was first isolated in its pure form in 1844. 1773-74 was the time when Persephone entered Aries, and Persephone went stationary-direct at 29 degrees Pisces within days of the date of the Boston *Tea* Party, an episode that eventually led to the American Declaration of

Independence. Persephone entered Taurus in 1805, the year that *morphine* was isolated. In 1882, as Persephone entered Cancer, the first law was enacted (in the U.S. and the world) requiring temperance education in public schools.

The late 1930s, with Persephone and Pluto in conjunction and ingressing together into Leo, was a particularly fertile period concerning drugs. In 1936 the Pan-American *Coffee* Bureau was organized, and during its first four years of advertising (1938-1941), U.S. coffee consumption was increased by 20 percent. In 1937 the *Marijuana* Tax Act was enacted. Last, but by no means least, in 1938 Dr. Albert Hofmann synthesized *LSD*.

The late 1960s and early 1970s, which we have shown in a previous chapter to be a potent time for Persephone-Pluto, constituted a highly significant and concentrated period for ceremonial chemistry. Public awareness of drugs was high (pun intended). *Tobacco* and *alcohol* use were at an all-time peak, as were *barbituates, marijuana, and LSD*. Laws were enacted to restrict or outlaw their use, and commissions for prevention and abuse were created. Nixon declared drug abuse as "America's Public Enemy No. 1." Poppy cultivation and *opium* production was forbidden in Turkey.

Not to be forgotten is the chemistry of birth control. Although various kinds of birth control pills had been experimented with since the mid-1950s, public awareness and widespread use really accelerated around 1970. *The Pill* was a drug not intended to alter consciousness, but it indirectly yet forcefully altered attitudes about relationship and brought together issues ruled by Venus, Persephone, Pluto, and Ceres: sex, equality, mothering, birth control (the abandonment of pregnancy), the freedom for women to pursue maturity as individuals and as sexual beings. As Venus, the antithesis of Demeter-mothering, stationed in Scorpio in 1970, all the known planets (including Ceres) were in the signs Virgo, Libra, Scorpio, or Taurus. We observe, once again, that Persephone is the hidden (unknown) linking mechanism that connects and integrates these signs, their ruling planets, and the issues they represent — all for the purpose of growth and awareness.

Chapter 18

PERSEPHONE, THE MYTH AND THE PLANET: CAN CONCLUSIONS BE MADE ABOUT A NEVER-ENDING STORY?

One of the biggest differences between our modern viewpoint and that of the ancients is that the latter were apparently quite at ease with the innate contradictions, duplications, repetitions, mirrorings, and "double standards" of their myths, while we moderns seem to be incapable of simply accepting duality and "letting it be." We cannot sit at peace in the presence of a paradox. On the contrary, we feel it is necessary, even "our duty," to resolve such inconsistencies and be able to state one final, irrefutable conclusion. Indeed, it is due to this modern habit that the reader probably expects this book, according to the norm, to have the obligatory concluding remarks. How is this possible, when the intention of the author has been to align with the spirit of the Persephone myth, the epitome of paradox?

We look to the modern treatment of the "gender problem" in the Persephone myth to illustrate our point. Marriage (inaugurated by rape, as it is in most myths) inevitably separates mother and daughter, who grieve for each other and long for reunion with the company of women. Men in general, and fathers specifically, are depicted in myth as the agency for such a separation. In the modern process of attempting to "resolve dualities," this has been interpreted to mean that marriage or relationship must be some kind of compromise (meaning that one or both participants must give up something), because it is seen as a contest of gain-and-loss. In the equation of separation = loss (of the overblown modern concept of individuality), *good* (innocence) and *evil* (the violator) are pitted against each other. Using this same thesis-antithesis logic, the myth has been further interpreted to mean "man-technology-patriarchy is the problem, and woman-nature-feminism is the solution." The kind of thinking that leads to such a conclusion is characteristic of the modern tendency to "divide and conquer," to clean up discrepancies, to conceptualize in terms of duality, which we assume must be by definition contradictory, and then to assign positive or negative value to either side, between which we must then *choose*.

Needless to say, this is not what the myth instructs us to do. Demeter may be enraged at Jupiter for permitting the abduction of Persephone, but never does she suggest that Pluto himself is an unworthy partner, or that the state of marriage and the relationship of the sexes is doomed to be a battlefield. The dilemma simply exists, and the "solution" (ask any Libran) is not to choose one solution, but to have both —- *some of each*. Ceres-Demeter and Pluto get to have some of Persephone all the time, or all of Persephone some of the time, but nobody gets to have all of her all of the time. And Persephone gets some of each as well; she lives in both worlds, the only character in Greek mythology to have that honor. The mechanism by which the "both" solution is attained is the *cyclic return*.

"Life is full of infinite absurdities, which, strangely enough, do not even need to appear plausible, since they are true."
— *Luigi Pirandello*

Mytho-psychologists and feminists have pounced on the Persephone myth with their well-meaning intentions to clear it of contradictions, and thus transform it into a model whereby we might find the one true answer to our modern dilemmas. But part of *our dilemma is that we demand consistency*, both as a requirement for good literature (myth) and as part of the psycho-social definition of sanity. This prejudice has placed many a Uranian and Gemini personality under institutional supervision, but it has also denied or devalued the contribution of the Libran types — the ones who can't decide because they can see both sides of an issue which they know to be inseparable, like two sides of a coin.

Astronomers are likewise attempting to resolve the inconsistencies between observation and theoretical formulas by identifying one and only one planetary body Transpluto as the source of perturbation of the other planets. Astrologers are not immune from this tendency, for many wish to settle, once and for all, the contradictions inherent in multiple planetary rulership and the inconsistent muddle of what Zip Dobyns called "alphabet soup." What all these modern views have in common — and where they all fail — whether the method be psychology, science, astrology, mythology, archaeology, or sociology (feminism), is in the attempt at control. The desire for consistency is nothing less than the ego's way to individualize, identify and label, and consequently to believe it *knows*. Understanding is a form of mental control. We fool ourselves into thinking that, if we eliminate the disturbing duplicities, the double meanings, the whole paradoxical and mirroring effect of life, we will have the answer.

It might be said that the author of this book, in the enthusiasm and conviction over defining Transpluto as Persephone, has committed this same error. What we hope has come across, however, is the more enduring message: that in order to grow in our awareness of our personal selves and of what life is about, we must learn to see that everything is connected, inter-related, and reflective; that the cycle is eternal and the discovery is on-going. *Et cetera.*

•••••••••••••••

"We used to think that if we knew one, we knew two, because *one and one are two*. We are finding that we must learn a great deal more about and."
                                        — *Sir Arthur Eddington*

Appendix

GLOSSARY AND NOTES ON ETYMOLOGY: THE MEANING OF NAMES IN THE PERSEPHONE MYTH AND ASSOCIATED STARS AND CONSTELLATIONS

Persephone was called the "Destroyer." She is considerably older than the Eleusinian myth found in classical writings. There is some evidence that she was really another name for Hecate (the Crone of the Triple Goddess Demeter-Persephone-Hecate), and had ruled the underworld as Destroying Mother Kali ruled it under the name of *Prisni,* which may have been the origin of Persephone's Etruscan name *Persipnei.* The Romans called her *Prosperpina.* It was under this name that she passed into Christian tradition as a Queen of She-Demons. Like Kali the Destroyer, she was the basic Death-goddess from the beginning.[262]

Athenians called Persephone *Pherrephata,* which means killer of suckling pigs.

Persephone has also been identified with the constellation Corona Borealis, the Northern Crown. This may have come about through a Chaldean title for this star group, *Persephon,* taken from *Phe'er* (Crown) and *Serphon* (Northern). If Charles Francois Dupuis, an 18th century writer responsible for this hypothesis, is correct, the origin of the constellational figure as well as the name may date back before Cretan times. Malphelcarre (one derivation of the Arabic name for this constellation), translates as "Circle of the Pupil of the Eye," but *pupillais* is the Latin equivalent of the Greek *Kore.*[263] Corona Borealis has gone by several associated names, including *Libera,* derived from *Liber Bacchus.*

Liber was an Italian god of fructification, a member of the oldest cycle of Roman gods, who became attached to Bacchus. Primarily he appears to have been a god of impregnation of both plants and animals. In 496 B.C. there was a severe famine in Rome, and in response, Demeter, Dionysos and Kore were imported and Latinized by the names Ceres, Liber, and Libera. A temple was erected and dedicated to these deities. Libera was considered the consort of Liber. Their festival, the Liberalia, was held in the spring. *Dis* was the Latinized form of the Greek Pluto, and the consort of Dis was *Proserpina,* identical with Libera. A cult of Dis and Proserpina was first brought to Rome about 249 B.C.[264]

The constellation *Libra:* Sometimes this was seen as a chariot black as night and pulled by coal-black horses. This chariot belonged to Hades who used it when he abducted Persephone from the adjacent constellation Virgo. The pictorial emblem of Libra may represent the beam of a balance in equilibrium or an altar. A beam with a feather on each side was used by the Egyptians to represent the equal weights of the Scales of Justice; the Egyptians had an instrument like this scale that measured the inundation of the Nile.[265]

The constellation *Virgo:* The story of the harvest or wheat-field maiden appeared in England and Scotland, where she was known as *Kernababy* (corn-baby). An image with a female shape was made of wheat, adorned with flowers and carried as a charm to a kern supper celebrating the harvest. All through the winter this figure was treasured, and in the spring she was tossed into a stream. The configuration of Virgo was often portrayed as a winged maiden holding in her left hand an ear of wheat or corn. In some illustrations she held a caduceus or scales. On certain Zodiacs from ancient Egypt she was portrayed as Isis holding the spica or her infant son Horus.[266]

Recalling that one of the names of Persephone is *Kore,* and that the harvest was her special festival, we give here the etymology of some words having the root "cor."[267]

Corn is the Old English name for the cereal grain wheat, in Scotland for oats. The ultimate origin of the word oat is uncertain, but Old English *ate* is related to *ath,* meaning oath or promise to God. *Corban* (Hebrew quorban) is an offering to God as a vow.

Core was seed in Middle English, hence *coriander*. *Kernel* is the diminutive of corn seed. *Corm* is the bulblike enlargement, the underground stem of flowers like the gladiolus; *cormus* Latin; *Kormos* Greek for tree trunk. *Cormophyte,* an old division of plants having roots.

Kor is a Hebrew measure; *koruna, krone* is *crown* (a measurement of English money); *corona* is Latin for crown. *Corolla,* the diminutive of Latin corona, refers to a garland; a *corollary* is an inference or deduction, from the Latin *corollarium,* originally the money paid for a garland. A *coroner* is a public official "officer of the crown," who decides the cause of death.

Corner comes from the Latin word meaning horn or point or spike. *Kern* is a printing term referring to corner. Greek *Keras* is horn; *keratin* is the name for the ingredient of horns, nails or claws. *Cornu* is Latin for horn, hence the *cornucopia* (horn of plenty). A *cornet* is a horn instrument. The *cornea* of the eye also originates from the root meaning horn.

The name *Demeter* has been thought by different translators to mean corn-land, or land under corn, a measure of allotted land, homesteads, family units.[268]

Iambe-Baubo: The poetic meter originally used by the Greek satiric poets was iambic, named for Iambe. The Greek word *iambein* meant "to assail verbally." Aristotle wrote in POETICS that "the meter called iambic was used for insults."

REFERENCES

1. For examples, the reader is referred to : "Objections to Astrology," *The Humanist* (1975), Vol. 35, No. 5; the writings of Carl Sagan; and the goals and activities of the Committee for the Scientific Investigation of Claims of the Paranormal (CSICOP), which believes that astrology is dangerous. One of CSICOP's latest campaigns is to urge newspapers to print disclaimers along with their Sun-sign astrology columns... the SCIence COPs' version of the surgeon general's warning (see "Astrology Alert," *Omni,* Jan. 1994, p. 81; and Cathy Corzine, "PSI-COPS Without a Clue," *The Mountain Astrologer,* Feb. 1994, pp. 27-30).

2. Barbara Koval, *The Lively Circle: Astrology and Science in Our Modern Age.* Astro Computing Services (1981), p. 105.

3. Ibid., p. 21.

4. For an excellent outline of an anarchistic theory of knowledge, see Paul Feyerabend, *Against Method.* London: NLB (1975). This philosopher of science argues that science is not sacrosanct, and that "there is no idea, however ancient and absurd, that is not capable of improving our knowledge."

5. See "The Search for a Planet Beyond Neptune," by Morton Grosser, *Isis* (1964), Vol. 55, pp. 163-183.

6. When Thomas C. Van Flandern of the U.S.Naval Observatory announced a renewal of the trans-Neptunian planet search in 1981, he used the term for several reasons. The new search was a renewal of the original efforts of Percival Lowell to find a planet of significant mass beyond Neptune, and there was still debate about Pluto's planetoid size as well as its inability (due to small mass) to account for irregularities in orbits of the known outer planets. Also, starting in 1979, Pluto's elliptical orbit brought it inside Neptune's orbit where it will be until 1999. At least temporarily, Pluto is not "trans-Neptunian." (Kendrick Frazier, "A Planet Beyond Pluto," *Mosaic,* Sept./Oct. 1981, p.28).

7. Exact discovery time indicated by Clyde Tombaugh's reminiscences in "The Discovery of Pluto: Some Generally Unknown Aspects of the Story," *Mercury* (May/June 1986), pp. 66-72.

8. See Kendrick Frazier, "A Planet Beyond Pluto," *Mosaic* (Sept./Oct. 1981), pp. 27-32.

9. The story of this astrology-astronomy collaboration is related fully in John Hawkins, *Transpluto: Or Should We Call Him Bacchus?*(1976). The graphic ephemeris was published as Theodore Landscheidt and H. Hausman, *Transpluto*, Aalen, Germany: Ebertin-Verlag (1972).

10. Published by Hawkins Enterprising Publications, Dallas, Texas (1976). The ephemeris was programmed by Neil Michelsen and developed from calculations by the astronomer Emile Sevin, with the assistance of astrologer Larry Ely, author of Ephemeridis Persephone.

11. Alert readers will note that this is opposed the Pluto position of the July 4, 1776 birthchart of the United States.

12. Like many other scientific ideas, the interest in pursuing astronomical evidence for Transpluto has gone in and out of fashion over the years. Recent articles (written in layman's language) include the following: "Hunting Planet X" (*Science News*, Vol. 132, No. 2, 7/11/1987, p.21); "The Search for Planet X" (*Newsweek*, Vol. 110, No. 2, 7/13/1987, p. 55); "The Pull of Planet X" (*Astronomy*, Vol. 16, No.8, Aug/1988, pp. 30-39); "Living with Uncertainty" (*Astronomy*, Aug/1988, p. 6); "Astronomy Express: Where's Planet X?" (*Sky & Telescope*, Vol. 78, No. 6, Dec/1989, pp. 596-599); "New Clues Point to an Elusive 10th Planet" (*Boston Globe*, 1/15/1990, p.3); "Lookout for Planet X" (*Ad Astra*, Vol. 2, No. 3, Mar/1990, p. 5); "Astronomers Renew Search for an Elusive Planet" (*Atlanta Journal*, 4/22/1990, sec G, p. 3); "The Hunt for Planet X" (*New Scientist*, Vol. 128, No. 1748-49, 12/22/1990, pp. 34-37); "Plates, Pluto, and Planets X" (*Sky & Telescope*, Vol. 81, No. 4, Apr/1991, pp. 360-361); "Focal Point: Worlds Apart" (*Sky & Telescope*, Vol. 82, No. 4, Oct/1991, pp. 340-341); "Say Goodbye to the Tenth Planet" (*New Scientist*, Vol. 132, No. 1797, 11/30/1991, p. 21); "Planet X: Going, Going...but not Quite Gone" (*Science*, Vol. 254, No. 5037, 12/6/1991, pp. 1454-1455); "The Planet that Came in from the Cold" (*New Scientist*, Vol. 136, No. 1847, 11/14/1992, pp. 24-25); "Science Times: Evidence for Planet X Evaporates in Spotlight of New Research" (*New York Times*, 6/1/1993, sec C, p. 8); "Planet X" (*Natural History*, Vol. 102, No. 6, Jun/1993, pp. 62-63; "Planet X is Dead" (*Discover*, Sept. 1993, p. 33); "No Planet X" (*Sky &*

Telescope, July 1993, p. 8); "Planet X: a Myth Exposed" (*Nature,* 5/6/1993, pp. 18-19); "Hopes Fade in Hunt for Planet X" (*New Scientist,* 1/30/1993, p. 18); "A New Member of the Family, *Astronomy* (Dec., 1992, pp. 38-39); "Planet X" (*Ad Astra,* Sept/Oct, 1993, pp. 47+.

13. Mark Littman, *Planets Beyond: Discovering the Outer Solar System.* John Wiley & Sons (1988), pp. 206 and 247.

14. Owen Gingerich, "The Naming of Uranus and Neptune," *Astronomical Society of the Pacific* leaflet No. 352 (October, 1958).

15. quoted in Steve Strimer's *The Astrology of Relativity,* Northampton MA: Aldebaran Press (1994). Evangeline Adams is known as the author of this statement, but she evidently lifted the words from Raphael.

16. William Graves Hoyt, *Planets X and Pluto* (Univ. of Arizona Press), p. 7.

17. Appropo of this is the book by Isabel Hickey, *Pluto or Minerva: The Choice is Yours* (1977) in which the author attempted to find a more positive meaning for Pluto

18. William Graves Hoyt, *Planets X and Pluto* (Univ. of Arizona Press), p. 215. Allusions to the Underworld or Death were common in discussions by astronomers over the possible name. One suggestion was "Posthumous," refering to the fact that the planet was discovered after Percival Lowell died. The influential director of Harvard Observatory, Harlow Shapely, wondered, "What kind of symbol would you use for the planet — a skull and crossbones?" (B.Z. Jones and L.G. Boyd, *The Harvard College Observatory*, Belknap, 1971, p. 330.)

19. A.C.D. Crommelin, "The Discovery of Pluto," *Monthly Notices of the Royal Astronomical Society* (1931), Vol. 91, p. 385.

20. Patrick Moore, "The Naming of Pluto," *Sky & Telescope* (Nov. 1984), pp.400-401.

21. Letter dated March 27, 1930 from trustee Roger Putnam to Slipher; quoted in William Graves (!) Hoyt, "William H. Pickering's Planetary Predictions and the Discovery of Pluto," *Isis* (1976), Vol. 67, # 239, p. 564.

22. Readers may be under the impression that nuclear power did not enter into general consciousness until 1945, when the world was

startled into a sudden realization by the destruction of Hiroshima. Actually, prior to W.W. II, there was much discussion and debate over the possibility of nuclear power. Some of the scientists that were directly involved in the creation of the atomic bomb had been inspired by the science fiction of H.G. Wells, such as *The World Set Free* (1914) which had predicted atomic bombs, and *The Shape of Things to Come* (1933). The subject was well known enough that "splitting the atom" had become a catch phrase during the 1920s. The early 1930s were a particularly vital time of scientific discoveries that led to the making of the bomb, and thousands of Jewish scientists and intellectuals were leaving Germany due to Hitler's regime. Ideas about nuclear fission were being shared openly among scientists throughout the world, regardless of politics and nationality, right up until the late 1930s. With the outbreak of war in Europe in 1939, the development of the atomic bomb then became shrouded in government secrecy. But the chain reaction for an atomic fission bomb had been so obvious an idea that it was recognized by newspaper writer William Laurence (author of *Men and Atoms*). His article in the *Saturday Evening Post* for September 7, 1940 was the last public comment on the possibility of an atomic bomb. (See Richard Rhodes, *The Making of the Atomic Bomb,* Simon & Schuster, 1986; and McGeorge Bundy, *Danger and Survival,* Vintage, 1988).

23. Astrophotography was one of the earliest applications of photography. On Jan. 7, 1839 came the announcement of Daguerre's discovery; within months astronomers were investigating the astronomical applications, and they began experimenting with methods like the change of focus; within a few years came the first attempts to photograph the Moon, a solar eclipse, and sunspots. See Dorrit Hoffleit, *Some Firsts in Astronomical Photography* (Harvard College Observatory, 1950).

24. Many qualities which Demetra George assigns to Ceres in her *Asteroid Goddesses* (ACS Publications, 1986) actually belong to Persephone. Eleanor Bach demonstrated the differences between two similar planets that are easily confused, the Moon and Ceres, in her article "Ceres," *The Mountain Astrologer* (Gem/Cancer 1989), pp. 7 ff. The misunderstood nature of Pluto was likewise discussed in the context of the Persephone

myth by Renee Parsons in "The Myths of Pluto," *The Mountain Astrologer* (Apr/May 1991), pp. 26-28.

25. Andrew Bevan perpetuates the mistaken name/myth of Bacchus in "The Discovery of Transpluto," *The Mountain Astrologer* (Dec./Jan. 1990), pp.31-33.

26. See C. Kerenyi, *Dionysos,* as well as his *Eleusis: Archetypal Image of Mother and Daughter,* Pantheon Books; Walter F. Otto, *Dionysus: Myth and Cult,* Indiana Univ. Press; Marcel Detienne, *Dionysos Slain,* Johns Hopkins Univ. Press; Arthur Evans, *The God of Ecstasy,* St. Martin's Press; Helene Deutsch, *A Psychoanalytic Study of the Myth of Dionysus and Apollo,* International Universities Press.

27. Elizabeth Kubler-Ross was responsible for developing this concept.

28. A recent addition to the onslaught of books and media promising to improve your relationships has an amusing title and author name: *Hot Monogamy* by Dr. Pat Love (yes, that is her real name), Sounds True Audio-Video.

29. From the Associated Press (July 6, 1993): The State of World Population report by the United Nations Population Fund states that the unprecedented growth in migration "could become the human crisis of our age." People are moving from rural areas to cities on a scale unknown in history, with more than 100 million migrants leaving their homes and pushing into territories occupied by others.

30. Gunther Zuntz, *Persephone: Three Essays on Religion and Thought in Magna Graecia* (Oxford: Clarendon Press, 1971), p.13.

31. Demetra George, p. 72 of "An Interview with Demetra George and Vicki Noble," *Welcome to Planet Earth* (Vol. 11, #10)

32. Charlene Spretnak, *Lost Goddesses of Early Greece,* Boston: Beacon Press (1984), p. 107.

33. New Age media is inundated (indeed, drowning) with these buzzwords of suffering-sacrifice, which tend to lose their significance with overuse. An example is Barbara Duffin's "Chiron and the Moon Goddess: The Shadow Myth," *The Mountain Astrologer* (March 1994), pp. 71-75, where the word "wound" or "wounded" appears no fewer than 14 times in the

final three paragraphs. This article is incidentally also an example of the current general confusion around correlation of myth and astrology (astrological Chiron is assigned characteristics that belong to the mythological complex of Persephone-Pluto-Ceres-Dionysos). Amplified by the enthusiasm of backlash feminism, this author mis-identifies Chiron as the mythological figure who can reconcile the masculine and feminine "wounding." "Wounded" is simply an image of how people have projected their own victim/lack/loss consciousness upon the gods themselves.

34. Some children's books include Penelope Farmer, *The Story of Persephone*, N.Y.: William Morrow (1973); *Demeter and Persephone*, translated by Penelope Proddow and illustrated by Barbara Cooney, Doubleday (1972); *Daughter of Earth: A Roman Myth*, retold and illustrated by Gerald McDermott, N.Y.: Delacorte (1984); Kris Waldherr, *Persephone and the Pomegranate*, N.Y.: Dial (1993); Cynthia Birrer, *Song to Demeter*, Lothrop, Lee & Shepard (1987); Claudia Fregosi, *Snow Maiden*, Prentice Hall (1979); Sarah Tomaino, *Persephone, Bringer of Spring*, N.Y.:Crowell (1971); Margaret Hodges, *Persephone and the Springtime*, Little, Brown (1973).

35. Jean Shinolda Bolen, *Goddesses in Everywoman*, N.Y.: Harper & Row (1985).

36. Jennifer Woolger and Roger Woolger, *The Goddess Within: A Guide to the Eternal Myths That Shape Women's Lives*, (Ballantine Books, 1989).

37. An adjective commonly attached by the Greeks to the names of characters in their mythology, such as fleet-footed Mercury, Pallas Athena, golden Venus, rosy-fingered Aurora, etc.

38. Paul Friedrich, *The Meaning of Aphrodite* (Univ. of Chicago Press), p. 25.

39. The Mesoamerican view of Venus is explained in Bruce Scofield, *Signs of Time: An Introduction to Mesoamerican Astrology*, Amherst MA: One Reed Publications (1994). The Babylonian omen tablets called the Enuma Anu Enlil, or "Venus Tablets of Ammizaduga," are the best known source that reveal the Mesopotamian fears about the destructive powers of the planet Venus. There is an abundance of literature debating the astronomical value of these tablets, but little serious

examination of the omens (except for writings by Velikovsky and his followers). See also Tim Johnson's "The Truth about Venus: an Expose," *The Mountain Astrologer*(Gem/Cancer 1989), p. 33, who reminds us that "the primary function of Venus is gratification. Not cooperation, not compromise, not balance, — just basic gratification."

40. See Richard Hinckley Allen, *Star Names: Their Lore and Meaning* (Dover, 1963), pp. 460 ff. for Virgo and 269 ff. for Libra. See also the writings of Alice Bailey, *Esoteric Astrology.* Ovid referred to Scorpio as a "double constellation." What we now call Libra was known to the Greeks as Chelae (the Claws).

41. Mircea Eliade, *The Forge and the Crucible* (Univ. of Chicago Press, 2nd edition), p. 38.

42. Ibid., p. 41.

43. An interesting socio-biological view of the paleolithic beginnings of love (Venus) and War (Mars) is given in Dudley Young, *Origins of the Sacred: The Ecstasies of Love and War,* N.Y.: St. Martin's Press.

44. John Hawkins, *Transpluto: Or Should We Call Him Bacchus, Ruler of Taurus?*

45. This is perfectly illustrated by the recent appearance of the pop psychology book, *Men Are from Mars, Women are from Venus: A Practical Guide for Improving Communication and Getting What You Want in Your Relationships,* by John Gray (HarperCollins, 1992). While this book is useful in identifying the differences between men and women, that's as far as it goes. The author encourages people to "understand" and manipulate the differences (in order to *gain* more and avoid *loss),*but there is no awareness of the recognition that we are the same, that men can learn to be/feel like women, and women can learn to think like men.

46. Richard Hinckley Allen, *Star Names: Their Lore and Meaning* (Dover, 1963), p. 363.

47. The system presented here conforms to much of the analysis in Jonathan Dunn's "Astronomical Clues to a Complete Rulership System," *Aspects* (Spring 1990), pp. 13-14. Dunn explains how we might expect there to be astronomical similarities between Pluto and the "missing" ruler of Libra, which he suggests *might* be a yet undiscovered trans-Plutonian body. Our system here

accounts for the gaps that often appear in rulerships tables as the *exaltation* and *fall* of the mutable signs (see "Rulerships and Dignities," part 6 of *The Mountain Astrologer's* Beginner's Guide to Astrology Series, *TMA*, Aug/Sept 1993, p.38), and the tables of planetary dignities and debilities in these publications: Rob Hand, *Horoscope Symbols,* Para Research (1981), p. 201; Michael Meyer, *A Handbook for the Humanistic Astrologer,* Anchor (1974), p. 79; Manly P. Hall, *Astrological Keywords,* Littlefield, Adams (1975), p. 83; *Alan Oken's Complete Astrology,* Bantam (1980), p. 311.

48. David Kinsley, *The Goddesses' Mirror* (State Univ. of N.Y., 1989) is an example of the glorification of Aphrodite and the confusion many researchers have over assigning to Venus the powers and purposes which myths clearly state belong to Persephone.

49. The medieval and Renaissance astrologers called this pattern the domiciles.

50. See Rob Baker, "The Dance of Libra," *Parabola* (Winter, 1991), pp. 20-27.

51. One very interesting approach to planetary pairing was given by Ron Alan Pierce in "The Music of the HemiSpheres: How Astrology Corresponds to Left-Brain and Right-Brain Processes," *The Mountain Astrologer* (March 1992), pp. 45, 54. The author discusses dual-rulership and pairs up the planets with regard to Spatial (right hemisphere) and Temporal (left hemisphere) Realms. His system can be extended to pair up the asteroids with Mercury, and Persephone as a match with Venus.

52. Richard Hinckley Allen, *Star Names: Their Lore and Meaning* (Dover, 1963), p. 468.

53. There is much evidence of the splitting and re-gendering of goddesses and their associated planets. The Akkadians originally worshipped the Sun as a female goddess, while the Moon and Venus were male deities. As the planet Venus came to be associated with Ishtar, she retained a dual nature: as the Morning Star she was the goddess of war and was called the Male Ishtar; in the Evening she became the goddess of love, and was called the Female Ishtar. (See Rupert Gleadow, *The Origin of the Zodiac,* Castle Books, p. 152.)

54. See Sylvia Perrea, *Descent to the Goddess* (N.Y.: Inner City Books, 1981) and the excellent treatise on *Inanna: Queen of Heaven and Earth* by Diane Wolkstein and Samuel Noah Kramer (Harper & Row, 1983). Also, Judith Ochshorn, "Ishtar and Her Cult," in *The Book of the Goddess Past and Present: An Introduction to Her Religion* (ed. Carl Olson, N.Y.: Crossroad), and the chapters on Ishtar-Inanna and Isis in Anne Baring and Jules Cashford, *The Myth of the Goddess: Evolution of an Image* (London: Viking. 1991).

55. See Wolkstein's translation, *Inanna,* pp. 57-60.

56. Many of the ideas expressed in the first half of this chapter were first described by Brad Clark (see acknowledgment).

57. See Brad Clark, "Proving Astrology: Some Thoughts While Reposing on the Horns of a Dilemma," *NCGR Memberletter* (Feb. 1989).

58. Brad Clark, personal correspondence.

59. Brad Clark provided this analogy in a taped lecture he gave on Neptune.

60. See Rob Hand, *Essays on Astrology,* Whitford Press (1982), and his subsequent taped lectures on the cardinal points in the constellations.

61. Mary Greer, *Tarot for Your Self* (Newcastle, 1984), p. 217.

62. A. E. Thierens, *Astrology and the Tarot* (Newcastle, 1975), p. 81.

63. Gail Fairfield, *Choice Centered Tarot* (1984), p. 102.

64. Gerd Ziegler, *Tarot: Mirror of the Soul* (Samuel Weiser, 1988), p. 59.

65. Thomas Kuhn, in his *Structure of Scientific Revolutions,* University of Chicago (1970), observed that while science thrives on new ideas, orthodox scientific training is not well designed to produce the thinker who will easily discover a fresh approach. Kuhn noted that a certain blindness occurs among scientists because they think they know "the right answers." This makes it particularly difficult for them to analyze an older science on its own terms. It is a pathetic comment on the myopic vision of Establishment Science to note that Kuhn was severely criticized by the scientific community for making these brilliant observations.

66. Pickering, "A Statistical Investigation of Comets," *Annals of the Harvard College Observatory* (1911), Vol. 61, p. 368.

67. William Graves Hoyt, "William Henry Pickering's Planetary Predictions and the Discovery of Pluto," *Isis* 67 (1976), p. 554.

68. by Robert G. Aitken, *Astronomical Society of the Pacific* pamphlet No. 211 (Sept. 1946).

69. Notice the nearly identical treatment of the subject in another astronomical text: "Two young mathematical astronomers, unknown to each other, set themselves the problem of locating [the trans-Uranian planet], Leverrier in France and Adams in England. Adams finished first and sent his results to Airy, the Astronomer Royal. Airy was unaccountably slow in taking action, and ...[he] and others in England were severely criticized for their apathetic attitude in the matter. The anomalous motion of Uranus was an outstanding problem of the time, and it is hard to understand why Airy would have neglected any possible solution. *Astronomers are always receiving letters from cranks and crackpots explaining the mysteries of the universe, but it must have been immediately evident from Adams' letter that he was an accomplished mathematician whose findings deserved careful attention.*" (Robert Richardson, *The Star Lovers*, Macmillan, 1967, p. 189)

70. Eugene Wigner, *Symmetries and Reflections*. Indiana University Press (1967), p. 192.

71. Bernard D'Espagnat, "The Quantum Theory and Reality," *Scientific American* (Nov. 1979), p. 158.

72. David Bohm, *Wholeness and the Implicate Order,* London: Routledge & Kegan Paul (1980).

73. see Fritjof Capra, *The Tao of Physics,* Boulder: Shambhala (1975).

74. Immanuel Velikovsky is the best known example of such ostracism by the powerful scientific establishment, but many investigators have been similarly persecuted, denounced and discredited, despite their great popular support. See *Velikovsky and Establishment Science,* Kronos Press (1971); Donald Goldsmith, *Scientists Confront Velikovsky,* Cornell University Press (1977); *Velikovsky Reconsidered,* Doubleday (1975); Alfred de Grazia, et al., *The Velikovsky Affair,* University Books

(1967); Valerie Vaughan, *Science and Conscience* (1990).

75. In the 1960s science discovered that there was another side to a "black hole," called a "white hole." The black/white holes were not the same hole, yet they were connected. Modern physics shows that if we could move faster than light, we could enter a black hole and instantly pop out a white hole billions of light years away from entry. These black/white hole connections were named "worm holes." (see John Wheeler, *Geometrodynamcis,* N.Y. Academy Press, 1962)

76. This idea is developed in an article that discusses how the split between astrology and astronomy occurred during the Scientific Revolution. See Valerie Vaughan, "What Do Astrology and Modern Science Have in Common?" *NCGR Journal,* Vol. 13, No. 2 (1994).

77. The study of the behavior of scientists, their constructed institutions, and their politics is a fascinating area of knowledge called the *sociology of science.* The belief system or religion that assumes the *scientific method* and objective consciousness is the best or only valid way toward achieving knowledge is called *Scientism.* For a discussion of these topics, particularly in regard to scientists' rejection of astrology, see: Paul Feyerabend, "The Strange Case of Astrology" in *Science in a Free Society* (NLB, 1978), pp. 91-96; Edward James, "On Dismissing Astrology and Other Irrationalities," and Daniel Rothbart, "Demarcating Genuine Science from Pseudoscience," in *Philosophy of Science and the Occult* (State Univ. of N.Y., 1982); Peter Wright, "A Study in the Legitimisation of Knowledge: The 'Success' of Medicine and the 'Failure' of Astrology," in *On the Margins of Science: The Social Construction of Rejected Knowledge* (Univ. of Keele, 1979); *The Reception of Unconventional Science,* ed. Seymour Mauskopf (Amer. Assoc. for the Advancement of Science Symposium 25); John Burnham, *How Superstition Won and Science Lost* (Rutgers Univ. Press); Richard Olson, *Science Deified and Science Defied* (Univ. of Calif. Press, 1982); Paul Feyerabend, *Against Method* (NLB, 1975).

78. A few attempts have been made in this direction. Percy Seymour, author of *Astrology: The Evidence of Science* (Lennard Publishing, 1988), is a respected astronomer who has

proposed a scientific basis for astrology based on the effects of magnetism. Physicist Peter Roberts has investigated astrology seriously in *The Message of Astrology*. Hans J. Eysenck and D.K.B. Nias, two psychologists have done the same in *Astrology: Science or Superstition* (St. Martin's Press, 1982). All have been subject to criticism by their professional colleagues. See "Dr. Zodiac," *Omni* (Dec., 1989), pp. 60-72 and "Oh Ye of Too Much Faith" (*New Scientist,* 7/20/1991, pp. 41ff). Eysenck also astonished the scientific community by writing a respectful obituary of Michel Gauquelin.

79. See William Graves Hoyt, *Planets X and Pluto* (Univ. of Arizona Press), p. 6-7, and J.R. Hind, "Unnoticed Observatons of Neptune," *Monthly Notices of the Royal Astronomical Society,* Vol. 10 (1850), p. 42.

80. See J.R. Hind, "Unnoticed Observations of Neptune," *Monthly Notices of the Royal Astronomical Society,* Vol. 10 (1850), p. 42.

81. See Charles T. Kowal and Stillman Drake, "Galileo's Observations of Neptune," *Nature,* Vol. 287 (1980), pp. 311-313. See also follow-up letters to the editor, *Nature,* Vol. 290 (1981), pp. 164-165; and J.R. Hind, "On Lamont's Observations as a Fixed Star," *Monthly Notices of the Royal Astronomical Society,* Vol. 11 (1851), p. 11.

82. Ibid.

83. B. Peirce, "Investigation in the action of Neptune to Uranus," *Proceedings of the American Academy of Arts and Sciences,* Vol. 1 (1847), p. 65.

84. See Stanley L. Jaki, "The Titius Bode Law," *Sky & Telescope* (May 1972), pp. 280-281, and Laki's note on "The Original Formulation of the Titius-Bode Law," *Journal of the History of Astronomy,* Vol. 3 (1972), pp. 136-138. Other sources for the discussion in this chapter include: Agnes M. Clerke, *A Popular History of Astronomy During the Nineteenth Century,* London: Adam and Charles Black (1908); Eric G. Forbes, "Gauss and the Discovery of Ceres," *Journal of the History of Astronomy,* Vol. 2 (1971), pp. 195-199; William Graves Hoyt, *Planets X and Pluto,* Univ. of Arizona Press; Dorrit Hoffleit, *Some Firsts in Astronomical Photography,* Harvard College Observatory (1950).

85. William Graves Hoyt, *Planets X and Pluto* (Univ. of Arizona Press), pp. 25-26.

86. Letter by Von Zach quoted in Eric G. Forbes, "Gauss and the Discovery of Ceres," *Journal for the History of Astronomy,* Vol. 2 (1971), p. 195.

87. Eleanor Bach, *Ephemerides of the Asteroids,* N.Y.: Celestial Communications (1973), p. 4.

88. Valerie Vaughan, "Weaving Through the Story of Computers: Pallas Athena Looms Large," *NCGR Journal* (Spring, 1994), pp. 27-33.

89. Agnes M. Clerke, *A Popular History of Astronomy during the Nineteenth Century,* London: Adam and Charles Black (1908), p.76.

90. See M.W. Ovenden, "Bode's Law and the Missing Planet," *Nature,* Vol. 239 (October 27, 1972), pp. 508-509; and D. Ter Haar, "On the Origin of the Solar System," in *Annual Review of Astronomy and Astrophysics.*

91. A handy list of discovery dates for the first 100 asteroids discovered between 1801 and 1868 is published in Willy Ley's *Watchers of the Skies* (New York: Viking, 1966), pp. 510-511.

92. William Graves Hoyt, *Planets X and Pluto,* p. 31.

93. P.W. Gifford, letter to the editor, *Popular Astronomy,* Vol. 28 (1920), p. 251.

94. Forest Ray Moulton, *Celestial Mechanics,* Macmillan (1964), p. 430.

95. Benoit B. Mandelbrot "discovered" this and published his ideas in *The Fractal Geometry of Nature* (1983); see also Anne E. Beversdorf, "Fractal Progressions: Astrology, Fractals and Human Experience," *The Mountain Astrologer* (Aug/Sept. 1994), pp. 5-10.

96. The Foundation for the Study of Cycles (2600 Michelson Dr., Suite 1570, Irvine CA 92715) states as its purpose: "to foster, promote, coordinate, conduct, and publish scientific research ... in the rhythmic fluctuations in natural and social phenomena, and to function as a clearinghouse for scientists working in this area."

97. Adapted from the table in A.E. Roy and M.W. Ovenden, "On

the Occurrence of Commensurable Mean Motions in the Solar System," *Monthly Notices of the Royal Astronomical Society,* Vol. 114 (1954), pp. 232-241, and idem., "The Mirror Theorem," vol. 115 (1955), pp. 296-309.

98. R.A. Bureau and L. B. Craine, "Sunspots and Planetary Orbits, *Nature,* Vol. 228 (Dec. 5, 1970), p. 984. For an astrological discussion, see Neil Michelsen, *Tables of Planetary Phenomena* (ACS Publications, 1990), pp. 198-199.

99. See "Were Titius and Bode Right?" *Sky & Telescope* (April 1987), p. 371. These models also point to the likelihood of some fairly substantial planet originally existing between Uranus and Neptune.

100. For an excellent outline of this topic and its relevance for the Velikovsky argument, see Robert W. Bass, "Did Worlds Collide?" *Pensee,* Vol. 4 (Summer 1974), pp. 8-20, reprinted in *Strange Universe: A Sourcebook of Curious Astronomical Observations,* ed. William R.Corliss (Glen Arm, MD: The Sourcebook Project), Vol. A-1, pp. A1-46 - A1-66). At the time of publication of this article, Bass was Professor of Physics and Astronomy at Brigham Young University.

101. Venus is a case in point. The period of Venus is nearly equal to 8/13 of a year. Since Venus makes one trip around in the Sun in 8/13 of a year, the number of trips it makes in one year is one divided by 8/13, which is 13/8 or one and 5/8. Therefore, it gains 5/8 of a lap on the earth during one year. The number of years it takes to gain one full lap is one divided by 5/8. This result is 8/5 or 1.6 earth years, 584 days, the synodic period of Venus. This resonance, along with the remarkable anomaly of the retrograde rotation of Venus, could represent relics of a previous near-collision between Venus and the Earth (See C.J. Ransom, "How Stable is the Solar System?" *Pensee,* May, 1972, pp. 16-17, 35). Venus, as the ancient Mesoamericans knew, provides a wealth of proportional relationships. Even its rotation period (243 days) is in sync with its synodic period, and therefore in a ratio with the earth's orbit: 243 days multiplied by 12 is equal to 584 days multiplied by 5; 243 days multiplied by 3 is equal to 365 days multiplied by 2. As the eminent dynamical astronomer E.W. Brown stated in his retirement speech as President of the American Astronomical Society, *there*

is no quantitative reason known to celestial mechanics why Mars, Earth and Venus could not have nearly collided in the past.

102. Richard A. Kerr, "Pluto's Orbital Motion Looks Chaotic," *Science,* Vol. 240 (May 20, 1988), pp. 986-987; Gerald Jay Sussman and Jack Wisdom, "Numerical Evidence That the Motion of Pluto is Chaotic," *Science,* Vol. 241 (July 22, 1988), pp. 433-437; Richard A. Kerr, "Does Chaos Permeate the Solar System?" *Science,* Vol. 244 (April 14, 1989), pp. 144-145; Roger W. Sinnott, "Do Orbits Change in 100 Million Years?" *Sky & Telescope* (Aug. 1987), pp. 182-183; see also the entire issue of *Nature,* Vol. 308 (April 19, 1984), which features articles connecting the extinction of dinosaurs to the "catastrophe" of comet showers or collision with asteroids, and the evidence that mass extinctions of species and cometary bombardment of the Earth are *periodic.*

103. R.S. Harrington and T.C. Van Flandern, *Icarus,* Vol. 39 (1979), pp. 131-136.

104. Zechariah Sitchin, *The Twelfth Planet* (Book One of *The Earth Chronicles),* Avon (1978). Of particular relevance are the chapters 6 and 7, pp. 173-235. Elizabeth MacIlhaney and Barbara Hand Clow have also published articles outlining some of his ideas that pertain to astrology (See "The Forgotten Planet," *The Mountain Astrologer,* March 1994, pp. 5-12). Sitchin wrote a synopsis of his latest and "most astrological" book, *When Time Began* (Avon, 1993), in this same issue of *TMA,* pp. 13-18. Bruce Scofield retorts on Sitchin and cult archaeology in "The Uranian Observer" (*TMA,* April/May 1994), and reader response to the Sitchin article is published in *TMA* (June 1994), pp. 77-79. For a general summary of the astronomical theories, ancient and modern, concerned with the origin of the solar system, see *The Origin of the Solar System,* ed. by S.F. Dermott (John Wiley, 1976). This covers collision theory, capture theory, even creation myths (though of course the authors cannot resist making the usual dig at astrology and the equally predictable praise of the triumph of science over all other approaches).

105. George Biddell Airy, "An Account of some circumstances historically connected with the discovery of the planet exterior

to Uranus," *Monthly Notices of the Royal Astronomical Society* (1846), Vol. 7, p. 124.

106. "Sur la position actuelle de la planete situee au dela de Neptune," *Comptes Rendus* (1848), Vol. 27, p. 203.

107. *Comptes Rendus* (1848), Vol. 27, p. 209.

108. Benjamin Peirce, "On the law of vegetable growth and the periods of the planets," *Proceedings of the American Academy of Arts and Sciences* (1852), Vol. 2, p. 241.

109. "Theorie et tables du mouvement de Mercure," *Annuales de'Observatoire Imperial de Paris* (1859), Vol. 5, p.1 ff., 102-106.

110. See W. G. Hoyt, *Lowell and Mars,* Univ. of Arizona (1976).

111. "Preliminary account of a speculative and practical search for a trans-Neptunian planet," *American Journal of Science* (1880), Vol. 20, 3rd Series, p. 225-234.

112. *Astronomie Populaire* (Paris, 1879), p. 661. Translated version: *Popular Astronomy,* Appleton, N.Y. (1907), pp. 471-472.

113. "On an ultra-Neptunian planet," *Proceedings of the Royal Society of Edinburgh*(1880), Vol. 10, p. 636, and "On Comets and Ultra-Neptunian Planets, *The Observatory* (1880), Vol. 3, p. 427, 436.

114. Morton Grosser, "The Seach for a Planet Beyond Neptune," *Isis* (1964), Vol. 55, p. 169.

115. F.C. Vassart, *"The twelve constituent planets of our solar system, of which three are ultra-Neptunian"* (1881); Gabriel Dallet, "Note sur les planetes extremes de notre systeme solaire," *Revue Scientifique* (1882), Vol. 4, p. 80; Oskar Reichenbach, *Two Planets Beyond Neptune and the Motion of the Solar System: A Speculation* (London, 1875); David Todd, "Telescopic Search for the trans-Neptunian planet," *Proceedings of the American Academy of Arts and Sciences* (1886), Vol. 21, p. 228-243; Isaac Roberts, "Photographic search for a planet beyond the orbit of Neptune, *Monthly Notices of the Royal Astronomical Society* (1892), Vol. 52, p. 501.

116. "Planetes inconnues," *Bulletin de la Societe Astronomique de France* (1900), Vol. 14, p. 340-341.

117. "Contribution a la recherche des planetes stuees au dela de

l'orbite de Neptune, *Bulletin de la Societe Astronomique de France* (1901), Vol. 15, p. 266-271.

118. "Nouvelle contribution a la recherche d'une planete transneptunienne," *Bulletin de la Societe Astronomique de France* (1902), Vol. 16, p. 31-32.

119. "Sur les planetes telescopiques," *Bulletin de la Societe Astronomique de France* (1901), Vol. 15, p. 358-361.

120. *Bulletin de la Societe Astronomique* (1902 and 1907).

121. "A photographic Search for Planet O," *Annals of the Astronomical Observatory of Harvard College* (1911), Vol. 61, Pt. 3, p. 360-373; "A search for a planet beyond Neptune," *Annals* (1909), Vol. 61, Pt. 2, p. 113-162; "The Assumed Planet O beyond Neptune," *The Observatory* (1909), Vol. 32, p. 326-328; "Perturbation de Neptune et planete transneptunienne," *Bulletin de la Societe Astronomique de France* (1919), Vol. 33, p. 393-394; "The Transneptunian Planet," *Annals* (1927), pp. 49-59; "The next planet beyond Neptune," *Popular Astronomy* (1928), Vol. 36, pp. 143-165 and 218-221; "The Orbits of the comets of short period," *Popular Astronomy* (1928), Vol. 36, pp. 274-281; "The orbit of Uranus," *Popular Astronomy* (1928), Vol. 36, pp. 353-361; "Planet O," *Popular Astronomy* (1929), Vol. 37, pp. 135-138; "The Perturbations of Neptune," *Harvard College Observatory Circular* #215 (May 15, 1919); "The Assumed Planet Beyond Neptune," *Popular Astronomy* (1909), Vol. 17, pp. 545-547; "The Trans-Neptunian Planet," *Ann. Harv. Coll. Observ.* (1919), Vol. 82, pp. 49-59; "The TransNeptunian Planet," *Popular Astronomy* (1930), Vol. 38, pp. 285-294; "Planet P, its Orbit, Position and Magnitude, Planets S & T," *Popular Astronomy* (1931), Vol. 39, pp. 385-398; "The Three Other Planets Beyond Neptune," *Popular Astronomy* (1928), Vol. 36, pp. 417-424.

122. W.H. Pickering, "A Statistical Investigation of Cometary Orbits," *Ann. Harv. Coll. Observ.* (1911), Vol. 61, pp. 167-368.

123. "On the cause of the remarkable circularity of the orbits of the planets and satellites, and on the origin of the planetary system," *Astronomische Nachrichten* (1909), Vol. 180, p. 194. See also *Researches on the Evolution of Stellar Systems,* Lynn, Mass: Nichols (1910), p. 375-376.

124. "Contribution a la recherche des planetes ultra-neptuniennes," *Comptes Rendus* (1909), Vol. 148, pp. 754-758. See also "Tables nouvelles de mouvements d'Uranus et de Neptune," *Annales de l'Observatoire de Paris* (1909), p. 28.

125. "Le Planete transneptunienne," *Bulletin de la Societe Astronomique de France* (1914),Vol. 28, pp. 276-283.

126. "La planete transneptunienne," *L'Astronomie* (1914), Vol. 28, pp. 276-283.

127. *Memoirs of the Lowell Observatory* (1915), Vol. I, No. 1, pp. 4-5.

128. Martin Grosser, "The Search for a Planet Beyond Neptune," *Isis* (1964), Vol. 55, p.178.

129. "Der transneptunische planet," *Das Weltall* (1921), Vol. 21, pp. 113-115.

130. *Bulletin de la Societe Astronomique de France* (1923), Vol. 35, pp. 386-394, 437; and (1927), Vol. 41, p. 19.

131. Martin Grosser, "The Search for a Planet Beyond Neptune," *Isis* (1964), Vol. 55, p. 182.

132. "On the Predictions of transneptunian planets from the perturbations of Uranus," *Proceedings of the National Academy of Sciences* (1930), Vol. 16, pp. 364-371.

133. The exception is V. Kourganoff, who published a detailed analysis confirming that both Lowell and Pickering had indeed detected evidence of Pluto's existence and that the discovery was not an accident. See Gibson Reaves, "Kourganoff's contributions to the history of the discovery of Pluto," *Publications of the Astronomical Society of the Pacific* (1951), Vol. 63, pp. 49-60.

134. Quoted in Clyde Tombaugh, "The Discovery of Pluto: Some Generally Unknown Aspects of the Story, Part II," *Mercury* (July-Aug. 1986), p. 102.

135. Pickering, "Planet P, Its Orbit, Position and Magnitude," *Popular Astronomy,* Vol. 39 (1931), pp. 385-398.

136. Mark Littman, *Planets Beyond* (Wiley, 1988), p. 196.

137. See Mark Littman, *Planets Beyond* (Wiley, 1988).

138. *Comptes Rendus des Seances de l'Academie de Science,* tomes 222 (1/21/1946), pp. 220, 221; 223 (9/16/1946), pp. 469-472

and (10/24/1946), pp. 653-655; see also "Une planete transplutoniene, *Bulletin de la Societe Astronomique de France* (1946), Vol. 60, p. 188-189.

139. *Popular Astronomy,* Vol. 57 (1949), p. 176.

140. Kritzinger's explanations appear in *Nachrichtenblatt der Astronomische Zentralstelle* (1954, Vol. 8, p. 4; 1957, Vol. 11, p. 4; 1959, Vol. 13, p. 3) and *Orion* (1955), Vol. 4, p. 484.

141. See David W. Hughes, "Pluto: the Little-Known Planet," *Nature,* Vol. 266 (Mar.24, 1977), pp. 307-308.

142. Dane Rudhyar may have been one of the first to contemplate this idea as the "long sextile" in his *Birth Patterns for a New Humanity* (Netherlands: Servire-Wassenaar, 1969), pp. 74-75; Rudhyar also referred to Pluto's perihelion approach as the "fecundation of Neptunian ideals by the relentless activity of Plutonian factors," ibid., p. 67. This book was later published as *Astrological Timing* (1972).

143. Edward J. Gunn, "Another Planet?" *New Scientist* (Nov. 12, 1970), p. 345.

144. D. Rawlins, "Is there a Tenth Planet in the Solar System? *Nature,* Vol. 240 (12/22/1972), p. 457.

145. Hoyt, *Planets X and Pluto,* pp. 255-257 describes the argument.

146. Hoyt, ibid., p. 255.

147. Quoted in Mark Littman, "Where Is Planet X?" *Sky & Telescope* (Dec. 1989), p. 599.

148. Hoyt, *Planets X and Pluto,* p. 247.

149. See Robert Harrington and Thomas Van Flandern, "The Satellites of Neptune and the Origin of Pluto," *Icarus,* Vol. 39 (1979), pp. 131-136; and Robert and Betty Harrington, "The Discovery of Pluto's Moon," *Mercury,* Vol. 8 (Jan/Feb 1979), pp. 1-3, 6, 17; ibid., "Pluto: Still an Enigma After 50 Years," *Sky & Telescope,* Vol. 59 (June 1980), pp. 452-454.

150. See *Sky and Telescope* (May 1984), p. 406.

151. See "No Close Encounters," *Sky & Telescope* (Dec. 1988), pp. 603-604.

152. Mark Littman, *Planets Beyond,* pp. 199-203. See also Luis Alvarez, et al., "Extraterrestrial Cause for the Cretaceous-Tertiary Extinction," *Science,* Vol. 208 (June 6, 1980), pp. 1095-

1108, and the entire issue of *Nature,* Vol. 308 (April 19, 1984).

153. "Hunting Planet X: A Nothing That Counts," *Science News* (July 11, 1987), p.21; "The Search for Planet X," *Newsweek* (July 13, 1987), p. 55; "Uranus is Perturbed: Usual Suspect Rounded Up," *Discover* (Sept. 1987), pp. 16,20.

154. Mark Littman, *Planets Beyond,* pp. 205-206..

155. See Mark Littman, "Where Is Planet X?" *Sky & Telescope* (Dec. 1989), pp. 596-599; R.S. Gomes, "On the Problem of the Search for Planet X," *Icarus,* Vol. 80 (1989), pp. 334-343.

156. Mark Littman, *Planets Beyond,* p. 203.

157. This is a problem for all areas of research in astrology. See Valerie Vaughan, "Open Letter to the Astrological Community, *NCGR Memberletter* (Jan/Feb 1994), pp.2-4, and the reply by Donna Cunningham and further response by Valerie Vaughan in the *NCGR Memberletter* (Mar/Apr 1994), pp. 18 ff. These letters discuss possible solutions for better organization of astrological literature, not only for improving the ease, accuracy, and quality of research, but also for developing the status of astrology as a respectable profession.

158. *Recent Advances in Natal Astrology: A Critical Review 1900-1975,* compiled by Geoffrey Dean (London: The Astrological Association, 1977), p.241.

159. See R.A. Jacobson, *The Language of Uranian Astrology* (1975), and U. Rudolph, *Introduction to the Hamburg School of Astrology* (London: The Astrological Association, 1974).

160. For an outline and (now out-of-date) bibliography on the subject, see *Recent Advances in Natal Astrology,* pp. 241 ff. See also *American Astrology,* Vol. 24 (1957), No. 11, p. 7; Vol. 25 (1957), No. 1, p. 5; No. 3, p. 27; No. 6, p. 31; No. 11, p. 26; No. 12, p. 37; *In Search* (Winter-Spring 1961, #A, pp. 7-19; *Aquarian Astrology* (Spring 1973), p. 52; Otto Wims, "The Case of the Hypothetical Planets," *A.F.A. Bulletin* (1974), Vol. 36, No. 11.

161. Charles A. Jayne, *The Unknown Planets: With Ephemerides* (Monroe, NY: Astrological Bureau, 1974).

162. Jayne evidently reproduced a table contained in the Winter/Spring 1960-61 issue of *In Search.*

163. Jayne, *The Unknown Planets*, p. 15.

164. Maurice Wemyss, *The Wheel of Life: Or, Scientific Astrology*, Vol. III (London: L.N. Fowler, 1930), p. 154.

165. See "Larry Ely Claims He's Found a New Planet," *Daily Hampshire Gazette* (Northampton, Mass.), 10/5/1977, p. 21.

166. Theodore Landscheidt and H. Hausman, *Transpluto: Graphical Ephemeris of its Tropical Zodiac Position 1878-1987 Geocentric and Heliocentric,* Aalen, Germany: Ebertin-Verlag (1972).

167. Among the few astrologers who have discussed the identity of Transpluto as Persephone are Lindsay River and Sally Gillespie, *The Knot of Time: Astrology and the Female Experience* (Harper & Row, 1987), p. 176, 267.

168. See Virgil's *Georgics* I, 5-7; Sophocles' *Antigone,* 1119-1120; Pausanius VII, xx, 1,2; Homer's *Odyssey,* i, 226; and the *Hymn to the Basket of Ceres,* by Callimachus, 71.

169. Andrew Bevan, "The Discovery of Transpluto," *The Mountain Astrologer* (Dec/Jan 1990), pp. 31-33. In this article Bevan uses certain criteria common to the discovery charts of Uranus, Neptune, Pluto and Chiron to predict a projected discovery date for "Bacchus" as Nov. 3, 1995. It happens that all the conditions he supposes are necessary for planetary discovery happen to also occur on Nov.30-Dec. 1, 2033 ... *Moon in a water sign* (Pisces), *North Node in either Taurus or Libra* (for that date, it's Libra), *one of the traditional planets in Scorpio* (Mercury), *Venus in the early degrees of Sagittarius* (on that date, zero Sag), and *Chiron in either Taurus or Libra* (it's in Taurus).

170. The reader is referred to Hawkins' discussion of the mythology of Bacchus in his book *Transpluto*, pp. 16-31.

171. Quoted from Rob Hand's taped lecture "Symbols of the Father Complex."

172. Rouben Paul Gregorian, "Mythology and Its Relevance to Astrology," *NCGR Journal Geocosmic Magazine* (Spring 1991), pp.41-43.

173. Barbara Duffin, "Chiron and the Moon Goddess: the Shadow Myth," *The Mountain Astrologer* (March 1994), p. 72.

174. These examples come from the files of Brad Clark and Valerie Vaughan.

175. Collections of Greek myths that contain the myth of Demeter and Persephone, written for older children, include: Thomas Bullfinch, *The Age of Fable* (N.Y.: New American Library, 1962), pp. 85-91; Padriac Colum, *The Golden Fleece and the Heroes Who Lived Before Achilles* (N.Y.: Macmillan, 1921), pp. 61-72; Ingri and Edgar D'Aulaire, *Book of Greek Myths* (1962), pp. 58-63.

176. Both Hodges and Cooney have received the coveted Caldecott Medal for the best children's picture book of the year. Hodges won the first-place award in 1985, as well as the runner-up Caldecott Honor Book in 1965. Cooney won the first place in 1980. (See *The Newbery and Caldecott Awards: A Guide to the Medal and Honor Books*, Chicago: American Library Assoc., 1994.)

177. McDermott won the Caldecott Medal in 1975 and the Caldecott Honor Award in 1994.

178. For a detailed examination and comparison of Ovid's Persephone story and the Homeric Hymn to Demeter, see Stephen Hind, *The Metamorphosis of Persephone* (Cambridge Univ. Press, 1987).

179. Claudia Fregosi, *Snow Maiden* (Prentice Hall, 1979).

180. Neil Howe and William Strauss, "The New Generation Gap," *Atlantic Monthly* (Dec., 1992) pp. 80 - 81.

181. Dane Rudhyar, *Astrological Timing* (1972); Alexander Ruperti, *Cycles of Becoming*, C.R.C.S. Publications (1978); Steve Cozzi, *Generations and the Outer Planet Cycles*, A.F.A. (1986); Liz Greene, *The Outer Planets and Their Cycles: The Astrology of the Collective*, C.R.C.S. Publications (1983); Rob Hand, "Neptune-Pluto Cycle: The Pulse of Western Civilization," *NCGR Journal* (Autumn 1990), pp. 27-35; E. Alan Meece, "The Cycle of Revolution: Uranus-Pluto," *The Mountian Astrologer* (June 1993), pp.28ff.; Ken McRitchite, "The Pluto Millenium," *Geocosmic News* (Fall 1988), pp. 15-18; Robert Blaschke, "The Pluto in Leo Generation," *The Ascendant* (Fall/Winter 1992), pp. 8 ff.; Valerie Vaughan, "Decades, Demographics or Delineation: How is History Written?" *The Mountain Astrologer* (Nov. 1993), pp. 71-75.

182. See Bruce Scofield's *Signs of Time: An Introduction to Mesoamerican Astrology*, Amherst MA :One Reed Publications (1994).

183. See Daniel Boorstin, *The Discoverers*, Vintage Books (1985), p. 5 for the story of how the Greek historian Herodotus explained the Precessional Cycle.

184. See George Sarton, *A History of Science*, John Wiley (1964).

185. Some astrologers have noticed a 680-year cycle of triple conjunctions among the outer planets Saturn, Uranus and Neptune, which occurs in the following pattern: All 3 planets conjunct within a year; after 680 years, they conjunct in pairs over a 5-year period; after 171 years (1/4 of 680) they conjunct again in pairs over a 5-year period; after 680 years, the entire pattern repeats again. The most recent time this occurred was 1988-1993; the one previous was 1308 A.D. (See Terry Lamb, "Emergence of a New Age: The Conjunction of Saturn, Neptune and Uranus," *NCGR Memberletter*.) Kenneth Bowser has pointed out that Saturn, Uranus, Neptune and Pluto fall into patterns of concurrent aspects in "A New Perspective on the Outer Planets," *The Mountain Astrologer*, March 1992, pp. 19-23. Steve Cozzi gives useful tables plus a great discussion of the 170+ year-pattern of multiple conjunctions and ingresses of the outer planets; see *Generations and the Outer Planet Cycles* (A.F.A., 1986).

186. See *Science News* (April 1, 1989), p. 189.

187. Merlin Stone, *When God Was a Woman* (1976).

188. Rhys Carpenter, *Discontinuity in Greek Civilization* (1968), pp. x, 41.

189. Stephen Schneider and Randi Londer, *The Coevolution of Climate and Life*, p. 109.

190. Jehuda Neumann and R. Marcel Sgust, "Harvest Dates in Ancient Mesopotamia as Possible Indicators of Climatic Variations," *Climate Change* I (1978), pp. 238-252.

191. Reid Bryson and Thomas Murray, *Climates of Hunger* (1977).

192. Schneider, op.cit., p. 110.

193. The various writings of the climatologist Hubert H. Lamb make this clear.

194. Brooks, *Climate Through the Ages* (1949). Readers familiar with the writings of Immanuel Velikovsky will recognize the significance of the geophysical upheavals and climatic changes of the 15th and 8th centuries B.C. which are documented in his *Worlds in Collision, Ages in Chaos,* and *Earth in Upheaval.* Velikovsky proposed that the body we now call Venus was once a comet, made contact with the earth, changing orbits and causing all manner of earthly disruption at two times in the middle of the 2nd millennium B.C. The Chaldeans, by the way, did not observe Venus as one of the planets until about the 8th c. B.C., when Mars became suddenly a much more fearful god than previously. Velikovsky's evidence shows that the Earth had gone through another upheaval during the 8th/7th c., and that by 687 B.C. the planet Venus had finally established an orbit we know today. Despite the controversy over Velikovsky's theories, it is still true that many of the calendar systems in various cultures were greatly revised during these periods, with accompanying alterations in mythologies. Scientists have discovered many phenomena that confirm several of Velikovsky's theories, but he is usually never given credit by the established scientific community. As an example, see "Theory of Earth's Spin" *(New York Times,* 1/19/1993, Sec. C, p. 2), which discusses the theory that the direction of spin of the planets Earth, Mercury, Venus and Mars is due to a previous collision in our solar system. Comyns Beaumont is another catastrophist who claimed that there was a comet in the early 14th c. B.C. that caused catastrophic conditions on earth (see Robert Stephanos, "Catastrophists in Collision," *Fate,* March 1994, p. 67).

195. Nigel Calder, *The Weather Machine* (1974).

196. Hubert H. Lamb, "The Early Medieval Warm Epoch and its Sequel," *Paleogeography, Paleoclimatology, Paleoecology* I (1965), pp. 13-37.

197. Calder, op.cit.

198. Reproduced from a brochure advertising the Foundation for the Study of Cycles, 2600 Michelson Dr., Suite 1570, Irvine CA 92715.

199. Schneider, op.cit.

200. Hubert H. Lamb, "Volcanic Dust in Atmosphere, with

Chronology and Assessement of its Meteorological Significance," *Philosophical Transactions of the Royal Society of London* (1970), A266: pp. 425-533.

201. In addition to the Precessional pattern, there is also a 41,000-year obliquity cycle (the earth's axis tilts now at 23.5 degrees, but can vary over 41,000 years between 22.1 and 24.5 degrees). We might note that 41,000 years divided by 60 equals 683 years, the Persephone cycle. See J.D. Hays, et al., "Variations in the Earth's Orbit: Pacemaker of the Ice Ages," *Science* (Dec. 10, 1976), pp. 1121-1132; "Ice Ages Attributed to Orbit Changes," *Science News* (Dec. 4, 1976), p. 356; Michael Rampino, et al., "Can Rapid Climatic Change Cause Volcanic Eruptions?" *Science* (Nov. 16, 1979), pp. 826-829.

202. *Earth in Space* (March 1991), p. 3.

203. For a discussion of this, see *Science* (Aug. 19, 1983), pp. 737 ff.

204. See Colman McCarthy, "Making Our Own Disasters," *Washington Post* (Feb. 26, 1994).

205. Liz Greene wrote, "Venus is an extroverted planet, seeking her fulfillment through stimulation from, and union with, beloved objects ... quarreling and raging in battle for the sake of the passionate experience of life ... is part of Venus' outward-directed world." (*The Astrology of Fate*, Samuel Weiser, 1984, p. 65). See also Paul Friedrich, *The Meaning of Aphrodite* (Univ. of Chicago Press), pp. 78-79.

206. See ibid., p. 161.

207. Ibid, p. 202.

208. Helen Fisher, *Anatomy of Love* (Fawcett Columbine, 1992), is a good read on this controversial topic of the bio-chemical basis of human relationship.

209. Synodic and sidereal periods of Venus and the time patterns of Sun-Venus relationships bounded by greatest elongation, conjunction, etc., are not the same as the repetitive female biological cycles universally associated with the Moon-menses cycle. Women may ovulate in the middle of their menses cycle, but they do not become pregnant each time they ovulate. The exception seems to prove the rule: the purpose of Venus is *procreative.*There is one Venus time pattern that is of physiological significance to women: her *rotation,* which is

anomalously retrograde, takes 243 days, equal to 9 lunar months, one human pregnancy cycle. In this Venus-Moon relationship we see a link between the mothering aspect and the sexuality of women. The hormones ruled by Venus are different than those associated with the menstral cycle. Pheromones, the odor lures or "musk" that humans and animals exude, appear to trigger infatuation and lead to reproduction. There may be some cyclic association between Venus periods and the excretion of sex attractants in humans. One interesting correlation that sociobiologist Helen Fisher has found may be proof of such an idea. In *Anatomy of Love* she relates worldwide patterns of divorce and remarriage that follow a eight-year cycle (a great majority of divorces occur after four years of marriage, and a great number of re-marriages occur 4 years following the divorce). This double "four-year-itch" could correspond to the 8-year pattern of Venus-Sun conjunctions. See Bruce Scofield, *Signs of Time*, op.cit.

210. George Thompson, *The Prehistoric Aegean* (Lawrence & Wishart), p. 236-237.

211. For further discussion of the association of nine with lunar goddesses, see Karl Kerenyi, *Goddesses of Sun and Moon* (Spring Publications, 1979), pp. 72-74.

212. See John Alden Knight, *Moon Up, Moon Down* (Charles Scribner's Sons, 1942; revised in 1972 by Jacqueline Knight). The author explains why American sportsmen have for decades carried the annual edition of *Solunar Tables,* the booklet that tells the four times a day that fish and game become the most active.

213. Thompson, *The Prehistoric Aegean*, p. 236-237.

214. ibid., p.219.

215. ibid., p. 220.

216. See Robert Cowen, "Earthquakes Are Nature's Way of Keeping Planet Earth Livable," *Christian Science Monitor* (Jan. 28, 1994), p. 3.

217. James E. McDonald, *UFOs: Greatest Scientific Problem of Our Times,* lecture presented before the American Society of Newspaper Editors (April 22, 1967). For a complete list of McDonald's writings concerning his scientific research into

UFOs, see Valerie Vaughan, *Science and Conscience* (1990).

218. See Ted Bloecher, *Report on the UFO Wave of 1947* (1967).

219. *The Crop Circle Enigma: Grounding the Phenomenon in Science, Culture and Metaphysics,* ed. by Ralph Noyes (Gateway, 1990). The astronomer Gerald Hawkins, known for his work on Stonehenge, has argued that certain arrangements of these circles and rings have definite mathematical properties suggesting previously unknown theorems in Euclidean geometry (See Ivars Peterson, "Euclid's Crop Circles," *Science News,* Vol. 141, Feb. 1, 1992, pp. 76-77).

220. William Broad, "Science Times: Strange New Microbes Hint at a Vast Subterranean World, *New York Times* (Dec. 28, 1993).

221. Developing the symbol construction for Persephone was a joint effort of this author with Bradley V. Clark (see acknowledgement).

222. For discussion, see Jeff Mayo, *The Planets and Human Behavior,* London: Fowler (1972); The Astrologer's Handbook Series No. 4.

223. Bode was not only the one who suggested the name Uranus, he also proposed the symbol mentioned here, a circle with a dot inside (the sun) crowned with an upward-pointing arrow. The symbolism was clear: that which is outside the solar system. Bode's symbol has prevailed throughout the astronomy community and among European astrologers, but American astrologers and some British astronomers use the familiar "circle with an antenna on top," which was devised in France in 1784 (C.A. Lubbock, *The Herschel Chronicle,* Cambridge Univ. Press, 1933, p. 201).

224. The natural forms of Nature are rich in symbol and meaning. This topic is discussed and illustrated in Theodor Schwenk, *Sensitive Chaos* (Schocken Books, 1976); Peter Stevens, *Patterns in Nature* (Little, Brown, 1974); and the entire issue of *Orion: Nature Quarterly* (Vol. 4, No. 1, Winter 1985).

225. Brad Clark is responsible for this idea of transforming Libra's symbol into Persephone's.

226. Jane Harrison, *Prolegomena to the Study of Greek Religion* (Meridian Books, 1960), pp. 150-151. Exceptions are the male mysteries such as the Orphic and Mithraic religions.

227. George Mylonas, *Eleusis and the Eleusinian Mysteries* (Princeton Univ. Press, 1961), p. 15.

228. Walter Friedrich Otto, p. 14 in *The Mysteries* Vol. 2 of Eranos-Jahrbuch, *Papers From the Eranos Yearbooks.*

229. Elinor Gadon, *The Once and Future Goddess* (Harper & Row, 1989), p. 144.

230. Elizabeth Kubler-Ross, *On Death and Dying* (Macmillan, 1969) reintroduced this idea that human beings could accept the great connection between grief and rage which appears at the time of death or separation.

231. It is the Orphic religion that called her slayer and dreaded. See M.P. Nilsson, *A History of Greek Religion* (Clarendon Press, 1949), p. 138, and Frederick Grant, *Hellenistic Religions* (Library of Religion, 1953), Vol. 2, p. 110.

232. See Erwin Rohde, *Psyche: The Cult of Souls and Belief in Immortality Among the Greeks* (Harper & Row, 1966), pp. 224-225.

233. Harrison, *Prolegomena*, p. 286.

234. Mylonas, *Eleusis*, p. 239.

235. Harrison, *Prolegomena,*, p. 136.

236. Mylonas, *Eleusis*, p. 243.

237. Mylonas, *Eleusis*, p. 244.

238. Here we have again the number nine. For a related discussion about the Nine Lords of the Night and the 9-days of the Venus-Sun inferior conjunction, see Bruce Scofield, *Signs of Time.*

239. Joan Chamberlain Engelsman, *The Feminine Dimension of the Divine*, Philadelphia: Westminster Press (1979), p.54.

240. Quoted in Engelsman, *The Feminine Dimension of the Divine*, p. 164.

241. Quoted in Liz Greene, *The Astrology of Fate* (Samuel Weiser, 1984), p. 218-219.

242. Mylonas, *Eleusis*, pp. 284-285; Cicero, *De Legibus*, 2,14,36.

243. Many of the ideas expressed in this section on calendars were developed in a comprehensive examination called *The Attic Festivals of Demeter and Their Relation to the Agricultural Year*, by Allaire Brumfield (Arno Press, 1981).

244. Marja Gimbutas, *The Goddesses and Gods of Old Europe* (Univ. of California Press, 1982), p. 214.

245. Ibid., p. 214.

246. Saloman Reinach, *Orpheus: A Study of Religion* (Horace Liveright, 1942), pp. 19-20; *Encyclopedia of Religion and Ethics* (ed. James Hastings, Edinburgh: T & T Clark, 1937), Vol. 12, p. 133. See also the view of anthropologist Marvin Harris in his *Cows, Pigs, Wars, and Witches: The Riddles of Culture* (Vintage, Random House, 1978) and *The Sacred Cow and the Abominable Pig: Riddles of Food and Culture* (Touchstone, Simon & Schuster, 1987).

247. Hans Biedermann, *Dictionary of Symbolism* (Facts on File, 1992), p. 265.

248. A.Leo Oppenheim, *Letters From Mesopotamia* (Univ. of Chicago Press, 1967), p. 45. See also V.L. Bullough, *Sexual Variance in Society and History* (Univ. of Chicago Press, 1976), p. 53.

249. Helen Fisher explains this view of the significance of the plow in her *Anatomy of Love* (Fawcett Columbine, 1992), p.106, and pp. 278-286.

250. Ibid., p. 279.

251. This subtitle is borrowed from an interesting essay by Marcel Detienne, "The Violence of Wellborn Ladies: Women in the Thesmophoria," pp. 129 ff. in *The Cuisine of Sacrifice Among the Greeks* (Univ. of Chicago Press, 1989).

252. See Maurice Olender, "Aspects of Baubo: Ancient Texts and Contexts," pp. 83 ff. in *Before Sexuality: The Construction of Erotic Experience in the Ancient Greek World,* ed. by David Halperin et al., Princeton Univ. Press.

253. These interpretations are itemized in George Mylonas, *Eleusis and the Eleusinian Mysteries,* pp. 296 ff.

254. Ibid., pp. 297-298.

255. The texts that describe this unveiling are listed in Olender's essay "Aspects of Baubo," op cit.

256. Again, these sources are identified in Olender's essay.

257. Ibid., p. 92

258. Ibid., pp. 98-99.

259. Beatrice Bruteau, "The Unknown Goddess," in *The Goddess Re-Awakening: The Feminine Principle Today,* compiled by Shirley Nicholson (Theosophical Publishing House, 1989), p. 71.

260. This section is based on their findings, published as *The Road to Eleusis: Unveiling the Secret of the Mysteries,* by R. Gordon Wasson, Albert Hofmann, and Carl Ruck (N.Y.: Harcourt, Brace, Jovanovich), and *Persephone's Quest: Entheogencs and the Origins of Religion,* by G. Gordon Wasson, Stella Kramrisch, Jonathan Ott, and Carl Ruck (Yale Univ. Press).

261. Data is taken from Thomas Szasz, *Ceremonial Chemistry: The Ritual Persecution of Drugs, Addicts, and Pushers* (Anchor Books, 1975).

262. Quoted from Barbara Walker, *The Women's Encyclopedia of Myths and Secrets* (Harper Row, 1983), p. 786.

263. Quoted from Gertrude and James Jobes, *Outer Space: Myths, Name Meanings, Calendars From the Emergence of History to the Present Day* (Scarecrow Press, 1964), p.155, 157.

264. Walter Addison Jayne, *The Healing Gods of Ancient Civilizations,* N.Y.: University Books, pp. 431, 459.

265. *Outer Space,* p. 202.

266. Ibid., p. 274.

267. All of the following terms can be found in a good dictionary.

268. Mylonas, *Eleusis and the Eleusinian Mysteries,* p. 50.

269. Published in *The Crop Circle Enigma,* ed. by Ralph Noyes (Gateway, 1990). This book was the most requested item on Interlibrary Loan in 1992 from Hampshire College, Amherst, Mass.

PERSEPHONE EPHEMERIS • 1900 - 2015

Monthly positions, Tropical Zodiac, Geocentric Longitude, 0 hrs. GMT.
Positions generated by Solar Fire from ASTROLABE.

JAN	1	1900	11°♋36'	JAN	1 1903	13°♋16'
FEB	1	1900	11°♋15'	FEB	1 1903	12°♋55'
MAR	1	1900	11°♋02'	MAR	1 1903	12°♋41'
APR	1	1900	10°♋57'	APR	1 1903	12°♋35'
MAY	1	1900	11°♋05'	MAY	1 1903	12°♋42'
JUN	1	1900	11°♋24'	JUN	1 1903	13°♋00'
JUL	1	1900	11°♋48'	JUL	1 1903	13°♋24'
AUG	1	1900	12°♋14'	AUG	1 1903	13°♋49'
SEP	1	1900	12°♋34'	SEP	1 1903	14°♋10'
OCT	1	1900	12°♋45'	OCT	1 1903	14°♋21'
NOV	1	1900	12°♋43'	NOV	1 1903	14°♋21'
DEC	1	1900	12°♋30'	DEC	1 1903	14°♋09'
JAN	1	1901	12°♋10'	JAN	1 1904	13°♋49'
FEB	1	1901	11°♋49'	FEB	1 1904	13°♋27'
MAR	1	1901	11°♋35'	MAR	1 1904	13°♋13'
APR	1	1901	11°♋30'	APR	1 1904	13°♋08'
MAY	1	1901	11°♋38'	MAY	1 1904	13°♋15'
JUN	1	1901	11°♋56'	JUN	1 1904	13°♋32'
JUL	1	1901	12°♋20'	JUL	1 1904	13°♋56'
AUG	1	1901	12°♋46'	AUG	1 1904	14°♋22'
SEP	1	1901	13°♋07'	SEP	1 1904	14°♋42'
OCT	1	1901	13°♋17'	OCT	1 1904	14°♋53'
NOV	1	1901	13°♋16'	NOV	1 1904	14°♋53'
DEC	1	1901	13°♋03'	DEC	1 1904	14°♋41'
JAN	1	1902	12°♋43'	JAN	1 1905	14°♋20'
FEB	1	1902	12°♋22'	FEB	1 1905	13°♋59'
MAR	1	1902	12°♋08'	MAR	1 1905	13°♋45'
APR	1	1902	12°♋03'	APR	1 1905	13°♋40'
MAY	1	1902	12°♋10'	MAY	1 1905	13°♋46'
JUN	1	1902	12°♋28'	JUN	1 1905	14°♋04'
JUL	1	1902	12°♋52'	JUL	1 1905	14°♋27'
AUG	1	1902	13°♋18'	AUG	1 1905	14°♋53'
SEP	1	1902	13°♋39'	SEP	1 1905	15°♋14'
OCT	1	1902	13°♋49'	OCT	1 1905	15°♋25'
NOV	1	1902	13°♋48'	NOV	1 1905	15°♋24'
DEC	1	1902	13°♋36'	DEC	1 1905	15°♋13'

JAN 1 1906	14°♋53'		JAN 1 1909	16°♋28'
FEB 1 1906	14°♋32'		FEB 1 1909	16°♋07'
MAR 1 1906	14°♋17'		MAR 1 1909	15°♋53'
APR 1 1906	14°♋12'		APR 1 1909	15°♋47'
MAY 1 1906	14°♋18'		MAY 1 1909	15°♋52'
JUN 1 1906	14°♋35'		JUN 1 1909	16°♋09'
JUL 1 1906	14°♋59'		JUL 1 1909	16°♋32'
AUG 1 1906	15°♋24'		AUG 1 1909	16°♋57'
SEP 1 1906	15°♋45'		SEP 1 1909	17°♋18'
OCT 1 1906	15°♋56'		OCT 1 1909	17°♋30'
NOV 1 1906	15°♋56'		NOV 1 1909	17°♋30'
DEC 1 1906	15°♋45'		DEC 1 1909	17°♋20'
JAN 1 1907	15°♋25'		JAN 1 1910	17°♋00'
FEB 1 1907	15°♋04'		FEB 1 1910	16°♋39'
MAR 1 1907	14°♋49'		MAR 1 1910	16°♋24'
APR 1 1907	14°♋43'		APR 1 1910	16°♋18'
MAY 1 1907	14°♋49'		MAY 1 1910	16°♋23'
JUN 1 1907	15°♋06'		JUN 1 1910	16°♋40'
JUL 1 1907	15°♋30'		JUL 1 1910	17°♋03'
AUG 1 1907	15°♋55'		AUG 1 1910	17°♋28'
SEP 1 1907	16°♋16'		SEP 1 1910	17°♋49'
OCT 1 1907	16°♋28'		OCT 1 1910	18°♋01'
NOV 1 1907	16°♋28'		NOV 1 1910	18°♋02'
DEC 1 1907	16°♋17'		DEC 1 1910	17°♋51'
JAN 1 1908	15°♋57'		JAN 1 1911	17°♋32'
FEB 1 1908	15°♋36'		FEB 1 1911	17°♋11'
MAR 1 1908	15°♋21'		MAR 1 1911	16°♋56'
APR 1 1908	15°♋15'		APR 1 1911	16°♋49'
MAY 1 1908	15°♋21'		MAY 1 1911	16°♋54'
JUN 1 1908	15°♋38'		JUN 1 1911	17°♋10'
JUL 1 1908	16°♋01'		JUL 1 1911	17°♋33'
AUG 1 1908	16°♋27'		AUG 1 1911	17°♋58'
SEP 1 1908	16°♋48'		SEP 1 1911	18°♋19'
OCT 1 1908	16°♋59'		OCT 1 1911	18°♋32'
NOV 1 1908	16°♋59'		NOV 1 1911	18°♋33'
DEC 1 1908	16°♋48'		DEC 1 1911	18°♋22'

Jan 1 1912	18°♋03'	Jan 1 1915	19°♋36'
Feb 1 1912	17°♋42'	Feb 1 1915	19°♋15'
Mar 1 1912	17°♋27'	Mar 1 1915	19°♋00'
Apr 1 1912	17°♋20'	Apr 1 1915	18°♋53'
May 1 1912	17°♋25'	May 1 1915	18°♋57'
Jun 1 1912	17°♋42'	Jun 1 1915	19°♋12'
Jul 1 1912	18°♋04'	Jul 1 1915	19°♋34'
Aug 1 1912	18°♋29'	Aug 1 1915	19°♋59'
Sep 1 1912	18°♋50'	Sep 1 1915	20°♋21'
Oct 1 1912	19°♋03'	Oct 1 1915	20°♋33'
Nov 1 1912	19°♋03'	Nov 1 1915	20°♋35'
Dec 1 1912	18°♋53'	Dec 1 1915	20°♋25'
Jan 1 1913	18°♋34'	Jan 1 1916	20°♋07'
Feb 1 1913	18°♋13'	Feb 1 1916	19°♋46'
Mar 1 1913	17°♋58'	Mar 1 1916	19°♋30'
Apr 1 1913	17°♋51'	Apr 1 1916	19°♋23'
May 1 1913	17°♋56'	May 1 1916	19°♋27'
Jun 1 1913	18°♋12'	Jun 1 1916	19°♋43'
Jul 1 1913	18°♋34'	Jul 1 1916	20°♋05'
Aug 1 1913	18°♋59'	Aug 1 1916	20°♋29'
Sep 1 1913	19°♋21'	Sep 1 1916	20°♋51'
Oct 1 1913	19°♋33'	Oct 1 1916	21°♋03'
Nov 1 1913	19°♋34'	Nov 1 1916	21°♋05'
Dec 1 1913	19°♋24'	Dec 1 1916	20°♋55'
Jan 1 1914	19°♋05'	Jan 1 1917	20°♋37'
Feb 1 1914	18°♋44'	Feb 1 1917	20°♋16'
Mar 1 1914	18°♋29'	Mar 1 1917	20°♋01'
Apr 1 1914	18°♋22'	Apr 1 1917	19°♋53'
May 1 1914	18°♋27'	May 1 1917	19°♋57'
Jun 1 1914	18°♋42'	Jun 1 1917	20°♋12'
Jul 1 1914	19°♋04'	Jul 1 1917	20°♋34'
Aug 1 1914	19°♋29'	Aug 1 1917	20°♋59'
Sep 1 1914	19°♋51'	Sep 1 1917	21°♋20'
Oct 1 1914	20°♋03'	Oct 1 1917	21°♋33'
Nov 1 1914	20°♋05'	Nov 1 1917	21°♋35'
Dec 1 1914	19°♋55'	Dec 1 1917	21°♋26'

Jan	1	1918	21°♋07'	Jan	1	1921	22°♋37'
Feb	1	1918	20°♋47'	Feb	1	1921	22°♋16'
Mar	1	1918	20°♋31'	Mar	1	1921	22°♋01'
Apr	1	1918	20°♋23'	Apr	1	1921	21°♋52'
May	1	1918	20°♋27'	May	1	1921	21°♋56'
Jun	1	1918	20°♋42'	Jun	1	1921	22°♋10'
Jul	1	1918	21°♋03'	Jul	1	1921	22°♋31'
Aug	1	1918	21°♋28'	Aug	1	1921	22°♋56'
Sep	1	1918	21°♋50'	Sep	1	1921	23°♋17'
Oct	1	1918	22°♋03'	Oct	1	1921	23°♋31'
Nov	1	1918	22°♋05'	Nov	1	1921	23°♋33'
Dec	1	1918	21°♋56'	Dec	1	1921	23°♋24'
Jan	1	1919	21°♋38'	Jan	1	1922	23°♋07'
Feb	1	1919	21°♋17'	Feb	1	1922	22°♋46'
Mar	1	1919	21°♋01'	Mar	1	1922	22°♋30'
Apr	1	1919	20°♋53'	Apr	1	1922	22°♋22'
May	1	1919	20°♋57'	May	1	1922	22°♋25'
Jun	1	1919	21°♋11'	Jun	1	1922	22°♋39'
Jul	1	1919	21°♋33'	Jul	1	1922	23°♋00'
Aug	1	1919	21°♋57'	Aug	1	1922	23°♋24'
Sep	1	1919	22°♋19'	Sep	1	1922	23°♋46'
Oct	1	1919	22°♋32'	Oct	1	1922	23°♋59'
Nov	1	1919	22°♋35'	Nov	1	1922	24°♋02'
Dec	1	1919	22°♋26'	Dec	1	1922	23°♋54'
Jan	1	1920	22°♋08'	Jan	1	1923	23°♋36'
Feb	1	1920	21°♋47'	Feb	1	1923	23°♋16'
Mar	1	1920	21°♋31'	Mar	1	1923	23°♋00'
Apr	1	1920	21°♋23'	Apr	1	1923	22°♋51'
May	1	1920	21°♋26'	May	1	1923	22°♋54'
Jun	1	1920	21°♋41'	Jun	1	1923	23°♋07'
Jul	1	1920	22°♋02'	Jul	1	1923	23°♋28'
Aug	1	1920	22°♋27'	Aug	1	1923	23°♋53'
Sep	1	1920	22°♋48'	Sep	1	1923	24°♋14'
Oct	1	1920	23°♋02'	Oct	1	1923	24°♋28'
Nov	1	1920	23°♋04'	Nov	1	1923	24°♋31'
Dec	1	1920	22°♋55'	Dec	1	1923	24°♋23'

Jan	1	1924	24°♋06'	Jan	1	1927	25°♋32'
Feb	1	1924	23°♋45'	Feb	1	1927	25°♋12'
Mar	1	1924	23°♋29'	Mar	1	1927	24°♋56'
Apr	1	1924	23°♋20'	Apr	1	1927	24°♋47'
May	1	1924	23°♋23'	May	1	1927	24°♋49'
Jun	1	1924	23°♋37'	Jun	1	1927	25°♋02'
Jul	1	1924	23°♋57'	Jul	1	1927	25°♋22'
Aug	1	1924	24°♋22'	Aug	1	1927	25°♋46'
Sep	1	1924	24°♋43'	Sep	1	1927	26°♋08'
Oct	1	1924	24°♋57'	Oct	1	1927	26°♋22'
Nov	1	1924	25°♋00'	Nov	1	1927	26°♋26'
Dec	1	1924	24°♋52'	Dec	1	1927	26°♋18'
Jan	1	1925	24°♋34'	Jan	1	1928	26°♋01'
Feb	1	1925	24°♋14'	Feb	1	1928	25°♋41'
Mar	1	1925	23°♋58'	Mar	1	1928	25°♋24'
Apr	1	1925	23°♋49'	Apr	1	1928	25°♋15'
May	1	1925	23°♋52'	May	1	1928	25°♋17'
Jun	1	1925	24°♋05'	Jun	1	1928	25°♋30'
Jul	1	1925	24°♋26'	Jul	1	1928	25°♋50'
Aug	1	1925	24°♋50'	Aug	1	1928	26°♋14'
Sep	1	1925	25°♋12'	Sep	1	1928	26°♋36'
Oct	1	1925	25°♋25'	Oct	1	1928	26°♋50'
Nov	1	1925	25°♋29'	Nov	1	1928	26°♋54'
Dec	1	1925	25°♋21'	Dec	1	1928	26°♋46'
Jan	1	1926	25°♋03'	Jan	1	1929	26°♋30'
Feb	1	1926	24°♋43'	Feb	1	1929	26°♋09'
Mar	1	1926	24°♋27'	Mar	1	1929	25°♋53'
Apr	1	1926	24°♋18'	Apr	1	1929	25°♋44'
May	1	1926	24°♋20'	May	1	1929	25°♋46'
Jun	1	1926	24°♋33'	Jun	1	1929	25°♋58'
Jul	1	1926	24°♋54'	Jul	1	1929	26°♋18'
Aug	1	1926	25°♋18'	Aug	1	1929	26°♋42'
Sep	1	1926	25°♋40'	Sep	1	1929	27°♋04'
Oct	1	1926	25°♋54'	Oct	1	1929	27°♋18'
Nov	1	1926	25°♋57'	Nov	1	1929	27°♋22'
Dec	1	1926	25°♋49'	Dec	1	1929	27°♋15'

JAN 1 1930	26°♋58'		JAN 1 1933	28°♋23'
FEB 1 1930	26°♋38'		FEB 1 1933	28°♋03'
MAR 1 1930	26°♋22'		MAR 1 1933	27°♋47'
APR 1 1930	26°♋12'		APR 1 1933	27°♋37'
MAY 1 1930	26°♋14'		MAY 1 1933	27°♋38'
JUN 1 1930	26°♋26'		JUN 1 1933	27°♋50'
JUL 1 1930	26°♋46'		JUL 1 1933	28°♋09'
AUG 1 1930	27°♋10'		AUG 1 1933	28°♋33'
SEP 1 1930	27°♋32'		SEP 1 1933	28°♋55'
OCT 1 1930	27°♋46'		OCT 1 1933	29°♋09'
NOV 1 1930	27°♋50'		NOV 1 1933	29°♋14'
DEC 1 1930	27°♋43'		DEC 1 1933	29°♋07'
JAN 1 1931	27°♋27'		JAN 1 1934	28°♋51'
FEB 1 1931	27°♋07'		FEB 1 1934	28°♋31'
MAR 1 1931	26°♋50'		MAR 1 1934	28°♋15'
APR 1 1931	26°♋41'		APR 1 1934	28°♋05'
MAY 1 1931	26°♋42'		MAY 1 1934	28°♋05'
JUN 1 1931	26°♋54'		JUN 1 1934	28°♋17'
JUL 1 1931	27°♋14'		JUL 1 1934	28°♋36'
AUG 1 1931	27°♋37'		AUG 1 1934	29°♋00'
SEP 1 1931	27°♋59'		SEP 1 1934	29°♋22'
OCT 1 1931	28°♋14'		OCT 1 1934	29°♋37'
NOV 1 1931	28°♋18'		NOV 1 1934	29°♋41'
DEC 1 1931	28°♋12'		DEC 1 1934	29°♋35'
JAN 1 1932	27°♋55'		JAN 1 1935	29°♋19'
FEB 1 1932	27°♋35'		FEB 1 1935	28°♋59'
MAR 1 1932	27°♋18'		MAR 1 1935	28°♋43'
APR 1 1932	27°♋09'		APR 1 1935	28°♋32'
MAY 1 1932	27°♋10'		MAY 1 1935	28°♋33'
JUN 1 1932	27°♋22'		JUN 1 1935	28°♋44'
JUL 1 1932	27°♋42'		JUL 1 1935	29°♋03'
AUG 1 1932	28°♋05'		AUG 1 1935	29°♋27'
SEP 1 1932	28°♋27'		SEP 1 1935	29°♋49'
OCT 1 1932	28°♋42'		OCT 1 1935	00°♌04'
NOV 1 1932	28°♋46'		NOV 1 1935	00°♌09'
DEC 1 1932	28°♋39'		DEC 1 1935	00°♌03'

JAN 1 1936	29°♋47'		JAN 1 1939	01°♌09'
FEB 1 1936	29°♋27'		FEB 1 1939	00°♌49'
MAR 1 1936	29°♋10'		MAR 1 1939	00°♌33'
APR 1 1936	29°♋00'		APR 1 1939	00°♌22'
MAY 1 1936	29°♋00'		MAY 1 1939	00°♌22'
JUN 1 1936	29°♋12'		JUN 1 1939	00°♌32'
JUL 1 1936	29°♋31'		JUL 1 1939	00°♌51'
AUG 1 1936	29°♋54'		AUG 1 1939	01°♌14'
SEP 1 1936	00°♌16'		SEP 1 1939	01°♌36'
OCT 1 1936	00°♌31'		OCT 1 1939	01°♌51'
NOV 1 1936	00°♌36'		NOV 1 1939	01°♌57'
DEC 1 1936	00°♌30'		DEC 1 1939	01°♌51'
JAN 1 1937	00°♌14'		JAN 1 1940	01°♌36'
FEB 1 1937	29°♋54'		FEB 1 1940	01°♌17'
MAR 1 1937	29°♋38'		MAR 1 1940	01°♌00'
APR 1 1937	29°♋27'		APR 1 1940	00°♌49'
MAY 1 1937	29°♋28'		MAY 1 1940	00°♌49'
JUN 1 1937	29°♋39'		JUN 1 1940	00°♌59'
JUL 1 1937	29°♋58'		JUL 1 1940	01°♌18'
AUG 1 1937	00°♌21'		AUG 1 1940	01°♌41'
SEP 1 1937	00°♌43'		SEP 1 1940	02°♌03'
OCT 1 1937	00°♌58'		OCT 1 1940	02°♌18'
NOV 1 1937	01°♌03'		NOV 1 1940	02°♌24'
DEC 1 1937	00°♌57'		DEC 1 1940	02°♌18'
JAN 1 1938	00°♌42'		JAN 1 1941	02°♌03'
FEB 1 1938	00°♌22'		FEB 1 1941	01°♌43'
MAR 1 1938	00°♌05'		MAR 1 1941	01°♌27'
APR 1 1938	29°♋55'		APR 1 1941	01°♌16'
MAY 1 1938	29°♋55'		MAY 1 1941	01°♌15'
JUN 1 1938	00°♌06'		JUN 1 1941	01°♌26'
JUL 1 1938	00°♌24'		JUL 1 1941	01°♌44'
AUG 1 1938	00°♌48'		AUG 1 1941	02°♌07'
SEP 1 1938	01°♌10'		SEP 1 1941	02°♌29'
OCT 1 1938	01°♌25'		OCT 1 1941	02°♌44'
NOV 1 1938	01°♌30'		NOV 1 1941	02°♌50'
DEC 1 1938	01°♌24'		DEC 1 1941	02°♌45'

241

JAN 1 1942	02°♌30'		JAN 1 1945	03°♌49'
FEB 1 1942	02°♌10'		FEB 1 1945	03°♌30'
MAR 1 1942	01°♌54'		MAR 1 1945	03°♌14'
APR 1 1942	01°♌43'		APR 1 1945	03°♌02'
MAY 1 1942	01°♌42'		MAY 1 1945	03°♌01'
JUN 1 1942	01°♌52'		JUN 1 1945	03°♌11'
JUL 1 1942	02°♌10'		JUL 1 1945	03°♌29'
AUG 1 1942	02°♌33'		AUG 1 1945	03°♌51'
SEP 1 1942	02°♌55'		SEP 1 1945	04°♌13'
OCT 1 1942	03°♌10'		OCT 1 1945	04°♌29'
NOV 1 1942	03°♌16'		NOV 1 1945	04°♌35'
DEC 1 1942	03°♌11'		DEC 1 1945	04°♌30'
JAN 1 1943	02°♌57'		JAN 1 1946	04°♌16'
FEB 1 1943	02°♌37'		FEB 1 1946	03°♌57'
MAR 1 1943	02°♌21'		MAR 1 1946	03°♌40'
APR 1 1943	02°♌09'		APR 1 1946	03°♌29'
MAY 1 1943	02°♌08'		MAY 1 1946	03°♌27'
JUN 1 1943	02°♌18'		JUN 1 1946	03°♌37'
JUL 1 1943	02°♌36'		JUL 1 1946	03°♌54'
AUG 1 1943	02°♌59'		AUG 1 1946	04°♌17'
SEP 1 1943	03°♌21'		SEP 1 1946	04°♌39'
OCT 1 1943	03°♌37'		OCT 1 1946	04°♌55'
NOV 1 1943	03°♌43'		NOV 1 1946	05°♌01'
DEC 1 1943	03°♌38'		DEC 1 1946	04°♌57'
JAN 1 1944	03°♌23'		JAN 1 1947	04°♌43'
FEB 1 1944	03°♌04'		FEB 1 1947	04°♌23'
MAR 1 1944	02°♌47'		MAR 1 1947	04°♌07'
APR 1 1944	02°♌36'		APR 1 1947	03°♌55'
MAY 1 1944	02°♌35'		MAY 1 1947	03°♌53'
JUN 1 1944	02°♌45'		JUN 1 1947	04°♌03'
JUL 1 1944	03°♌03'		JUL 1 1947	04°♌20'
AUG 1 1944	03°♌26'		AUG 1 1947	04°♌43'
SEP 1 1944	03°♌47'		SEP 1 1947	05°♌04'
OCT 1 1944	04°♌03'		OCT 1 1947	05°♌20'
NOV 1 1944	04°♌09'		NOV 1 1947	05°♌27'
DEC 1 1944	04°♌04'		DEC 1 1947	05°♌23'

Jan 1 1948	05°♌09'			Jan 1 1951	06°♌27'
Feb 1 1948	04°♌50'			Feb 1 1951	06°♌08'
Mar 1 1948	04°♌33'			Mar 1 1951	05°♌51'
Apr 1 1948	04°♌21'			Apr 1 1951	05°♌39'
May 1 1948	04°♌20'			May 1 1951	05°♌37'
Jun 1 1948	04°♌29'			Jun 1 1951	05°♌46'
Jul 1 1948	04°♌46'			Jul 1 1951	06°♌03'
Aug 1 1948	05°♌09'			Aug 1 1951	06°♌25'
Sep 1 1948	05°♌31'			Sep 1 1951	06°♌47'
Oct 1 1948	05°♌46'			Oct 1 1951	07°♌03'
Nov 1 1948	05°♌53'			Nov 1 1951	07°♌10'
Dec 1 1948	05°♌49'			Dec 1 1951	07°♌07'
Jan 1 1949	05°♌35'			Jan 1 1952	06°♌53'
Feb 1 1949	05°♌15'			Feb 1 1952	06°♌34'
Mar 1 1949	04°♌59'			Mar 1 1952	06°♌17'
Apr 1 1949	04°♌47'			Apr 1 1952	06°♌05'
May 1 1949	04°♌45'			May 1 1952	06°♌03'
Jun 1 1949	04°♌55'			Jun 1 1952	06°♌12'
Jul 1 1949	05°♌12'			Jul 1 1952	06°♌28'
Aug 1 1949	05°♌34'			Aug 1 1952	06°♌51'
Sep 1 1949	05°♌56'			Sep 1 1952	07°♌12'
Oct 1 1949	06°♌12'			Oct 1 1952	07°♌28'
Nov 1 1949	06°♌19'			Nov 1 1952	07°♌36'
Dec 1 1949	06°♌15'			Dec 1 1952	07°♌32'
Jan 1 1950	06°♌01'			Jan 1 1953	07°♌18'
Feb 1 1950	05°♌42'			Feb 1 1953	06°♌59'
Mar 1 1950	05°♌25'			Mar 1 1953	06°♌43'
Apr 1 1950	05°♌13'			Apr 1 1953	06°♌31'
May 1 1950	05°♌11'			May 1 1953	06°♌28'
Jun 1 1950	05°♌20'			Jun 1 1953	06°♌37'
Jul 1 1950	05°♌37'			Jul 1 1953	06°♌53'
Aug 1 1950	05°♌59'			Aug 1 1953	07°♌16'
Sep 1 1950	06°♌21'			Sep 1 1953	07°♌37'
Oct 1 1950	06°♌37'			Oct 1 1953	07°♌54'
Nov 1 1950	06°♌45'			Nov 1 1953	08°♌01'
Dec 1 1950	06°♌41'			Dec 1 1953	07°♌58'

JAN 1 1954	07°♌44'		JAN 1 1957	09°♌00'
FEB 1 1954	07°♌25'		FEB 1 1957	08°♌41'
MAR 1 1954	07°♌09'		MAR 1 1957	08°♌25'
APR 1 1954	06°♌56'		APR 1 1957	08°♌12'
MAY 1 1954	06°♌54'		MAY 1 1957	08°♌09'
JUN 1 1954	07°♌02'		JUN 1 1957	08°♌17'
JUL 1 1954	07°♌18'		JUL 1 1957	08°♌33'
AUG 1 1954	07°♌40'		AUG 1 1957	08°♌55'
SEP 1 1954	08°♌02'		SEP 1 1957	09°♌17'
OCT 1 1954	08°♌19'		OCT 1 1957	09°♌33'
NOV 1 1954	08°♌26'		NOV 1 1957	09°♌41'
DEC 1 1954	08°♌23'		DEC 1 1957	09°♌38'
JAN 1 1955	08°♌10'		JAN 1 1958	09°♌25'
FEB 1 1955	07°♌51'		FEB 1 1958	09°♌07'
MAR 1 1955	07°♌34'		MAR 1 1958	08°♌50'
APR 1 1955	07°♌22'		APR 1 1958	08°♌37'
MAY 1 1955	07°♌19'		MAY 1 1958	08°♌34'
JUN 1 1955	07°♌27'		JUN 1 1958	08°♌42'
JUL 1 1955	07°♌43'		JUL 1 1958	08°♌58'
AUG 1 1955	08°♌05'		AUG 1 1958	09°♌19'
SEP 1 1955	08°♌27'		SEP 1 1958	09°♌41'
OCT 1 1955	08°♌43'		OCT 1 1958	09°♌58'
NOV 1 1955	08°♌51'		NOV 1 1958	10°♌06'
DEC 1 1955	08°♌48'		DEC 1 1958	10°♌03'
JAN 1 1956	08°♌35'		JAN 1 1959	09°♌50'
FEB 1 1956	08°♌17'		FEB 1 1959	09°♌32'
MAR 1 1956	07°♌59'		MAR 1 1959	09°♌15'
APR 1 1956	07°♌47'		APR 1 1959	09°♌02'
MAY 1 1956	07°♌44'		MAY 1 1959	08°♌59'
JUN 1 1956	07°♌52'		JUN 1 1959	09°♌06'
JUL 1 1956	08°♌09'		JUL 1 1959	09°♌22'
AUG 1 1956	08°♌30'		AUG 1 1959	09°♌43'
SEP 1 1956	08°♌52'		SEP 1 1959	10°♌05'
OCT 1 1956	09°♌09'		OCT 1 1959	10°♌22'
NOV 1 1956	09°♌16'		NOV 1 1959	10°♌30'
DEC 1 1956	09°♌13'		DEC 1 1959	10°♌28'

Jan 1 1960	10°♌15'		Jan 1 1963	11°♌29'
Feb 1 1960	09°♌57'		Feb 1 1963	11°♌11'
Mar 1 1960	09°♌40'		Mar 1 1963	10°♌54'
Apr 1 1960	09°♌27'		Apr 1 1963	10°♌41'
May 1 1960	09°♌24'		May 1 1963	10°♌37'
Jun 1 1960	09°♌31'		Jun 1 1963	10°♌44'
Jul 1 1960	09°♌47'		Jul 1 1963	10°♌59'
Aug 1 1960	10°♌08'		Aug 1 1963	11°♌20'
Sep 1 1960	10°♌30'		Sep 1 1963	11°♌42'
Oct 1 1960	10°♌47'		Oct 1 1963	11°♌59'
Nov 1 1960	10°♌55'		Nov 1 1963	12°♌08'
Dec 1 1960	10°♌52'		Dec 1 1963	12°♌06'
Jan 1 1961	10°♌40'		Jan 1 1964	11°♌54'
Feb 1 1961	10°♌21'		Feb 1 1964	11°♌36'
Mar 1 1961	10°♌05'		Mar 1 1964	11°♌19'
Apr 1 1961	09°♌52'		Apr 1 1964	11°♌05'
May 1 1961	09°♌48'		May 1 1964	11°♌02'
Jun 1 1961	09°♌55'		Jun 1 1964	11°♌09'
Jul 1 1961	10°♌11'		Jul 1 1964	11°♌24'
Aug 1 1961	10°♌32'		Aug 1 1964	11°♌45'
Sep 1 1961	10°♌54'		Sep 1 1964	12°♌07'
Oct 1 1961	11°♌11'		Oct 1 1964	12°♌23'
Nov 1 1961	11°♌19'		Nov 1 1964	12°♌32'
Dec 1 1961	11°♌17'		Dec 1 1964	12°♌30'
Jan 1 1962	11°♌05'		Jan 1 1965	12°♌18'
Feb 1 1962	10°♌46'		Feb 1 1965	12°♌00'
Mar 1 1962	10°♌30'		Mar 1 1965	11°♌43'
Apr 1 1962	10°♌16'		Apr 1 1965	11°♌30'
May 1 1962	10°♌13'		May 1 1965	11°♌26'
Jun 1 1962	10°♌20'		Jun 1 1965	11°♌33'
Jul 1 1962	10°♌35'		Jul 1 1965	11°♌48'
Aug 1 1962	10°♌56'		Aug 1 1965	12°♌09'
Sep 1 1962	11°♌18'		Sep 1 1965	12°♌30'
Oct 1 1962	11°♌35'		Oct 1 1965	12°♌47'
Nov 1 1962	11°♌43'		Nov 1 1965	12°♌56'
Dec 1 1962	11°♌41'		Dec 1 1965	12°♌54'

Jan 1 1966	12°♌43'	Jan 1 1969	13°♌55'
Feb 1 1966	12°♌25'	Feb 1 1969	13°♌38'
Mar 1 1966	12°♌08'	Mar 1 1969	13°♌21'
Apr 1 1966	11°♌55'	Apr 1 1969	13°♌07'
May 1 1966	11°♌50'	May 1 1969	13°♌03'
Jun 1 1966	11°♌57'	Jun 1 1969	13°♌09'
Jul 1 1966	12°♌12'	Jul 1 1969	13°♌23'
Aug 1 1966	12°♌32'	Aug 1 1969	13°♌44'
Sep 1 1966	12°♌54'	Sep 1 1969	14°♌06'
Oct 1 1966	13°♌11'	Oct 1 1969	14°♌23'
Nov 1 1966	13°♌20'	Nov 1 1969	14°♌32'
Dec 1 1966	13°♌19'	Dec 1 1969	14°♌31'
Jan 1 1967	13°♌07'	Jan 1 1970	14°♌20'
Feb 1 1967	12°♌49'	Feb 1 1970	14°♌02'
Mar 1 1967	12°♌32'	Mar 1 1970	13°♌45'
Apr 1 1967	12°♌19'	Apr 1 1970	13°♌31'
May 1 1967	12°♌15'	May 1 1970	13°♌27'
Jun 1 1967	12°♌21'	Jun 1 1970	13°♌33'
Jul 1 1967	12°♌35'	Jul 1 1970	13°♌47'
Aug 1 1967	12°♌56'	Aug 1 1970	14°♌08'
Sep 1 1967	13°♌18'	Sep 1 1970	14°♌29'
Oct 1 1967	13°♌35'	Oct 1 1970	14°♌46'
Nov 1 1967	13°♌44'	Nov 1 1970	14°♌56'
Dec 1 1967	13°♌43'	Dec 1 1970	14°♌55'
Jan 1 1968	13°♌31'	Jan 1 1971	14°♌44'
Feb 1 1968	13°♌14'	Feb 1 1971	14°♌26'
Mar 1 1968	12°♌56'	Mar 1 1971	14°♌09'
Apr 1 1968	12°♌43'	Apr 1 1971	13°♌56'
May 1 1968	12°♌39'	May 1 1971	13°♌51'
Jun 1 1968	12°♌45'	Jun 1 1971	13°♌56'
Jul 1 1968	13°♌00'	Jul 1 1971	14°♌10'
Aug 1 1968	13°♌21'	Aug 1 1971	14°♌31'
Sep 1 1968	13°♌42'	Sep 1 1971	14°♌52'
Oct 1 1968	13°♌59'	Oct 1 1971	15°♌10'
Nov 1 1968	14°♌08'	Nov 1 1971	15°♌19'
Dec 1 1968	14°♌07'	Dec 1 1971	15°♌19'

Jan 1 1972	15°♌08'		Jan 1 1975	16°♌19'
Feb 1 1972	14°♌50'		Feb 1 1975	16°♌01'
Mar 1 1972	14°♌33'		Mar 1 1975	15°♌45'
Apr 1 1972	14°♌19'		Apr 1 1975	15°♌31'
May 1 1972	14°♌15'		May 1 1975	15°♌25'
Jun 1 1972	14°♌20'		Jun 1 1975	15°♌30'
Jul 1 1972	14°♌34'		Jul 1 1975	15°♌44'
Aug 1 1972	14°♌55'		Aug 1 1975	16°♌04'
Sep 1 1972	15°♌16'		Sep 1 1975	16°♌25'
Oct 1 1972	15°♌34'		Oct 1 1975	16°♌43'
Nov 1 1972	15°♌43'		Nov 1 1975	16°♌53'
Dec 1 1972	15°♌42'		Dec 1 1975	16°♌53'
Jan 1 1973	15°♌31'		Jan 1 1976	16°♌42'
Feb 1 1973	15°♌14'		Feb 1 1976	16°♌25'
Mar 1 1973	14°♌57'		Mar 1 1976	16°♌08'
Apr 1 1973	14°♌43'		Apr 1 1976	15°♌54'
May 1 1973	14°♌38'		May 1 1976	15°♌49'
Jun 1 1973	14°♌44'		Jun 1 1976	15°♌54'
Jul 1 1973	14°♌58'		Jul 1 1976	16°♌07'
Aug 1 1973	15°♌18'		Aug 1 1976	16°♌28'
Sep 1 1973	15°♌40'		Sep 1 1976	16°♌49'
Oct 1 1973	15°♌57'		Oct 1 1976	17°♌06'
Nov 1 1973	16°♌06'		Nov 1 1976	17°♌16'
Dec 1 1973	16°♌06'		Dec 1 1976	17°♌16'
Jan 1 1974	15°♌55'		Jan 1 1977	17°♌05'
Feb 1 1974	15°♌38'		Feb 1 1977	16°♌48'
Mar 1 1974	15°♌21'		Mar 1 1977	16°♌31'
Apr 1 1974	15°♌07'		Apr 1 1977	16°♌17'
May 1 1974	15°♌02'		May 1 1977	16°♌12'
Jun 1 1974	15°♌07'		Jun 1 1977	16°♌17'
Jul 1 1974	15°♌21'		Jul 1 1977	16°♌30'
Aug 1 1974	15°♌41'		Aug 1 1977	16°♌50'
Sep 1 1974	16°♌03'		Sep 1 1977	17°♌12'
Oct 1 1974	16°♌20'		Oct 1 1977	17°♌29'
Nov 1 1974	16°♌30'		Nov 1 1977	17°♌39'
Dec 1 1974	16°♌29'		Dec 1 1977	17°♌39'

Jan 1 1978	17°♌29'	
Feb 1 1978	17°♌12'	
Mar 1 1978	16°♌55'	
Apr 1 1978	16°♌41'	
May 1 1978	16°♌35'	
Jun 1 1978	16°♌40'	
Jul 1 1978	16°♌53'	
Aug 1 1978	17°♌13'	
Sep 1 1978	17°♌34'	
Oct 1 1978	17°♌52'	
Nov 1 1978	18°♌02'	
Dec 1 1978	18°♌02'	
Jan 1 1979	17°♌52'	
Feb 1 1979	17°♌35'	
Mar 1 1979	17°♌18'	
Apr 1 1979	17°♌04'	
May 1 1979	16°♌58'	
Jun 1 1979	17°♌03'	
Jul 1 1979	17°♌16'	
Aug 1 1979	17°♌35'	
Sep 1 1979	17°♌57'	
Oct 1 1979	18°♌14'	
Nov 1 1979	18°♌25'	
Dec 1 1979	18°♌25'	
Jan 1 1980	18°♌15'	
Feb 1 1980	17°♌58'	
Mar 1 1980	17°♌41'	
Apr 1 1980	17°♌27'	
May 1 1980	17°♌21'	
Jun 1 1980	17°♌26'	
Jul 1 1980	17°♌39'	
Aug 1 1980	17°♌59'	
Sep 1 1980	18°♌20'	
Oct 1 1980	18°♌37'	
Nov 1 1980	18°♌48'	
Dec 1 1980	18°♌48'	
Jan 1 1981	18°♌38'	
Feb 1 1981	18°♌21'	
Mar 1 1981	18°♌04'	
Apr 1 1981	17°♌50'	
May 1 1981	17°♌44'	
Jun 1 1981	17°♌48'	
Jul 1 1981	18°♌01'	
Aug 1 1981	18°♌21'	
Sep 1 1981	18°♌42'	
Oct 1 1981	19°♌00'	
Nov 1 1981	19°♌10'	
Dec 1 1981	19°♌11'	
Jan 1 1982	19°♌01'	
Feb 1 1982	18°♌44'	
Mar 1 1982	18°♌28'	
Apr 1 1982	18°♌13'	
May 1 1982	18°♌07'	
Jun 1 1982	18°♌11'	
Jul 1 1982	18°♌24'	
Aug 1 1982	18°♌44'	
Sep 1 1982	19°♌05'	
Oct 1 1982	19°♌22'	
Nov 1 1982	19°♌33'	
Dec 1 1982	19°♌34'	
Jan 1 1983	19°♌24'	
Feb 1 1983	19°♌07'	
Mar 1 1983	18°♌51'	
Apr 1 1983	18°♌36'	
May 1 1983	18°♌30'	
Jun 1 1983	18°♌34'	
Jul 1 1983	18°♌47'	
Aug 1 1983	19°♌06'	
Sep 1 1983	19°♌27'	
Oct 1 1983	19°♌45'	
Nov 1 1983	19°♌56'	
Dec 1 1983	19°♌56'	

JAN 1 1984	19°♌47'		JAN 1 1987	20°♌55'
FEB 1 1984	19°♌31'		FEB 1 1987	20°♌39'
MAR 1 1984	19°♌13'		MAR 1 1987	20°♌23'
APR 1 1984	18°♌59'		APR 1 1987	20°♌08'
MAY 1 1984	18°♌53'		MAY 1 1987	20°♌01'
JUN 1 1984	18°♌57'		JUN 1 1987	20°♌04'
JUL 1 1984	19°♌10'		JUL 1 1987	20°♌17'
AUG 1 1984	19°♌29'		AUG 1 1987	20°♌36'
SEP 1 1984	19°♌50'		SEP 1 1987	20°♌57'
OCT 1 1984	20°♌08'		OCT 1 1987	21°♌15'
NOV 1 1984	20°♌19'		NOV 1 1987	21°♌26'
DEC 1 1984	20°♌19'		DEC 1 1987	21°♌27'
JAN 1 1985	20°♌10'		JAN 1 1988	21°♌18'
FEB 1 1985	19°♌53'		FEB 1 1988	21°♌02'
MAR 1 1985	19°♌37'		MAR 1 1988	20°♌45'
APR 1 1985	19°♌22'		APR 1 1988	20°♌30'
MAY 1 1985	19°♌16'		MAY 1 1988	20°♌24'
JUN 1 1985	19°♌19'		JUN 1 1988	20°♌27'
JUL 1 1985	19°♌32'		JUL 1 1988	20°♌39'
AUG 1 1985	19°♌51'		AUG 1 1988	20°♌58'
SEP 1 1985	20°♌12'		SEP 1 1988	21°♌20'
OCT 1 1985	20°♌30'		OCT 1 1988	21°♌37'
NOV 1 1985	20°♌41'		NOV 1 1988	21°♌48'
DEC 1 1985	20°♌42'		DEC 1 1988	21°♌50'
JAN 1 1986	20°♌33'		JAN 1 1989	21°♌41'
FEB 1 1986	20°♌16'		FEB 1 1989	21°♌24'
MAR 1 1986	20°♌00'		MAR 1 1989	21°♌08'
APR 1 1986	19°♌45'		APR 1 1989	20°♌53'
MAY 1 1986	19°♌39'		MAY 1 1989	20°♌46'
JUN 1 1986	19°♌42'		JUN 1 1989	20°♌50'
JUL 1 1986	19°♌54'		JUL 1 1989	21°♌02'
AUG 1 1986	20°♌13'		AUG 1 1989	21°♌21'
SEP 1 1986	20°♌35'		SEP 1 1989	21°♌42'
OCT 1 1986	20°♌52'		OCT 1 1989	21°♌59'
NOV 1 1986	21°♌04'		NOV 1 1989	22°♌11'
DEC 1 1986	21°♌05'		DEC 1 1989	22°♌12'

JAN 1 1990	22°♌03'		JAN 1 1993	23°♌10'
FEB 1 1990	21°♌47'		FEB 1 1993	22°♌54'
MAR 1 1990	21°♌31'		MAR 1 1993	22°♌38'
APR 1 1990	21°♌16'		APR 1 1993	22°♌23'
MAY 1 1990	21°♌09'		MAY 1 1993	22°♌16'
JUN 1 1990	21°♌12'		JUN 1 1993	22°♌18'
JUL 1 1990	21°♌24'		JUL 1 1993	22°♌30'
AUG 1 1990	21°♌43'		AUG 1 1993	22°♌49'
SEP 1 1990	22°♌04'		SEP 1 1993	23°♌10'
OCT 1 1990	22°♌21'		OCT 1 1993	23°♌27'
NOV 1 1990	22°♌33'		NOV 1 1993	23°♌39'
DEC 1 1990	22°♌34'		DEC 1 1993	23°♌41'
JAN 1 1991	22°♌26'		JAN 1 1994	23°♌33'
FEB 1 1991	22°♌10'		FEB 1 1994	23°♌17'
MAR 1 1991	21°♌53'		MAR 1 1994	23°♌00'
APR 1 1991	21°♌38'		APR 1 1994	22°♌45'
MAY 1 1991	21°♌31'		MAY 1 1994	22°♌38'
JUN 1 1991	21°♌34'		JUN 1 1994	22°♌40'
JUL 1 1991	21°♌46'		JUL 1 1994	22°♌52'
AUG 1 1991	22°♌04'		AUG 1 1994	23°♌10'
SEP 1 1991	22°♌25'		SEP 1 1994	23°♌31'
OCT 1 1991	22°♌43'		OCT 1 1994	23°♌49'
NOV 1 1991	22°♌55'		NOV 1 1994	24°♌01'
DEC 1 1991	22°♌57'		DEC 1 1994	24°♌03'
JAN 1 1992	22°♌48'		JAN 1 1995	23°♌55'
FEB 1 1992	22°♌32'		FEB 1 1995	23°♌39'
MAR 1 1992	22°♌15'		MAR 1 1995	23°♌23'
APR 1 1992	22°♌00'		APR 1 1995	23°♌07'
MAY 1 1992	21°♌54'		MAY 1 1995	23°♌00'
JUN 1 1992	21°♌56'		JUN 1 1995	23°♌02'
JUL 1 1992	22°♌08'		JUL 1 1995	23°♌13'
AUG 1 1992	22°♌27'		AUG 1 1995	23°♌32'
SEP 1 1992	22°♌48'		SEP 1 1995	23°♌53'
OCT 1 1992	23°♌06'		OCT 1 1995	24°♌11'
NOV 1 1992	23°♌17'		NOV 1 1995	24°♌23'
DEC 1 1992	23°♌19'		DEC 1 1995	24°♌25'

JAN	1	1996	24°♌17'		
FEB	1	1996	24°♌01'		
MAR	1	1996	23°♌44'		
APR	1	1996	23°♌29'		
MAY	1	1996	23°♌22'		
JUN	1	1996	23°♌24'		
JUL	1	1996	23°♌35'		
AUG	1	1996	23°♌54'		
SEP	1	1996	24°♌15'		
OCT	1	1996	24°♌33'		
NOV	1	1996	24°♌44'		
DEC	1	1996	24°♌46'		

Month	Day	Year	Value		
JAN	1	1996	24°♌17'	JAN 1 1999	25°♌22'
FEB	1	1996	24°♌01'	FEB 1 1999	25°♌07'
MAR	1	1996	23°♌44'	MAR 1 1999	24°♌51'
APR	1	1996	23°♌29'	APR 1 1999	24°♌35'
MAY	1	1996	23°♌22'	MAY 1 1999	24°♌27'
JUN	1	1996	23°♌24'	JUN 1 1999	24°♌29'
JUL	1	1996	23°♌35'	JUL 1 1999	24°♌40'
AUG	1	1996	23°♌54'	AUG 1 1999	24°♌58'
SEP	1	1996	24°♌15'	SEP 1 1999	25°♌19'
OCT	1	1996	24°♌33'	OCT 1 1999	25°♌37'
NOV	1	1996	24°♌44'	NOV 1 1999	25°♌49'
DEC	1	1996	24°♌46'	DEC 1 1999	25°♌52'
JAN	1	1997	24°♌38'	JAN 1 2000	25°♌44'
FEB	1	1997	24°♌23'	FEB 1 2000	25°♌29'
MAR	1	1997	24°♌06'	MAR 1 2000	25°♌12'
APR	1	1997	23°♌51'	APR 1 2000	24°♌57'
MAY	1	1997	23°♌44'	MAY 1 2000	24°♌49'
JUN	1	1997	23°♌46'	JUN 1 2000	24°♌51'
JUL	1	1997	23°♌57'	JUL 1 2000	25°♌02'
AUG	1	1997	24°♌15'	AUG 1 2000	25°♌20'
SEP	1	1997	24°♌36'	SEP 1 2000	25°♌40'
OCT	1	1997	24°♌54'	OCT 1 2000	25°♌59'
NOV	1	1997	25°♌06'	NOV 1 2000	26°♌11'
DEC	1	1997	25°♌08'	DEC 1 2000	26°♌13'
JAN	1	1998	25°♌00'	JAN 1 2001	26°♌06'
FEB	1	1998	24°♌45'	FEB 1 2001	25°♌50'
MAR	1	1998	24°♌29'	MAR 1 2001	25°♌34'
APR	1	1998	24°♌13'	APR 1 2001	25°♌19'
MAY	1	1998	24°♌06'	MAY 1 2001	25°♌11'
JUN	1	1998	24°♌08'	JUN 1 2001	25°♌12'
JUL	1	1998	24°♌18'	JUL 1 2001	25°♌23'
AUG	1	1998	24°♌37'	AUG 1 2001	25°♌41'
SEP	1	1998	24°♌57'	SEP 1 2001	26°♌02'
OCT	1	1998	25°♌15'	OCT 1 2001	26°♌20'
NOV	1	1998	25°♌27'	NOV 1 2001	26°♌32'
DEC	1	1998	25°♌30'	DEC 1 2001	26°♌35'

Jan 1 2002	26°♌27'	Jan 1 2005	27°♌32'
Feb 1 2002	26°♌12'	Feb 1 2005	27°♌17'
Mar 1 2002	25°♌56'	Mar 1 2005	27°♌01'
Apr 1 2002	25°♌41'	Apr 1 2005	26°♌46'
May 1 2002	25°♌33'	May 1 2005	26°♌37'
Jun 1 2002	25°♌34'	Jun 1 2005	26°♌38'
Jul 1 2002	25°♌44'	Jul 1 2005	26°♌49'
Aug 1 2002	26°♌02'	Aug 1 2005	27°♌06'
Sep 1 2002	26°♌23'	Sep 1 2005	27°♌27'
Oct 1 2002	26°♌41'	Oct 1 2005	27°♌45'
Nov 1 2002	26°♌53'	Nov 1 2005	27°♌58'
Dec 1 2002	26°♌56'	Dec 1 2005	28°♌01'
Jan 1 2003	26°♌49'	Jan 1 2006	27°♌54'
Feb 1 2003	26°♌34'	Feb 1 2006	27°♌39'
Mar 1 2003	26°♌18'	Mar 1 2006	27°♌23'
Apr 1 2003	26°♌02'	Apr 1 2006	27°♌07'
May 1 2003	25°♌54'	May 1 2006	26°♌59'
Jun 1 2003	25°♌55'	Jun 1 2006	27°♌00'
Jul 1 2003	26°♌06'	Jul 1 2006	27°♌10'
Aug 1 2003	26°♌23'	Aug 1 2006	27°♌27'
Sep 1 2003	26°♌44'	Sep 1 2006	27°♌48'
Oct 1 2003	27°♌02'	Oct 1 2006	28°♌06'
Nov 1 2003	27°♌15'	Nov 1 2006	28°♌19'
Dec 1 2003	27°♌18'	Dec 1 2006	28°♌22'
Jan 1 2004	27°♌11'	Jan 1 2007	28°♌15'
Feb 1 2004	26°♌56'	Feb 1 2007	28°♌01'
Mar 1 2004	26°♌39'	Mar 1 2007	27°♌45'
Apr 1 2004	26°♌24'	Apr 1 2007	27°♌29'
May 1 2004	26°♌16'	May 1 2007	27°♌20'
Jun 1 2004	26°♌17'	Jun 1 2007	27°♌21'
Jul 1 2004	26°♌27'	Jul 1 2007	27°♌31'
Aug 1 2004	26°♌45'	Aug 1 2007	27°♌48'
Sep 1 2004	27°♌06'	Sep 1 2007	28°♌09'
Oct 1 2004	27°♌24'	Oct 1 2007	28°♌27'
Nov 1 2004	27°♌36'	Nov 1 2007	28°♌40'
Dec 1 2004	27°♌39'	Dec 1 2007	28°♌43'

Jan 1 2008	28°♌37'	Jan 1 2011	29°♌41'
Feb 1 2008	28°♌22'	Feb 1 2011	29°♌26'
Mar 1 2008	28°♌06'	Mar 1 2011	29°♌10'
Apr 1 2008	27°♌50'	Apr 1 2011	28°♌55'
May 1 2008	27°♌42'	May 1 2011	28°♌46'
Jun 1 2008	27°♌43'	Jun 1 2011	28°♌46'
Jul 1 2008	27°♌53'	Jul 1 2011	28°♌55'
Aug 1 2008	28°♌10'	Aug 1 2011	29°♌12'
Sep 1 2008	28°♌30'	Sep 1 2011	29°♌33'
Oct 1 2008	28°♌49'	Oct 1 2011	29°♌51'
Nov 1 2008	29°♌01'	Nov 1 2011	00°♍04'
Dec 1 2008	29°♌05'	Dec 1 2011	00°♍08'
Jan 1 2009	28°♌58'	Jan 1 2012	00°♍02'
Feb 1 2009	28°♌43'	Feb 1 2012	29°♌48'
Mar 1 2009	28°♌27'	Mar 1 2012	29°♌31'
Apr 1 2009	28°♌12'	Apr 1 2012	29°♌15'
May 1 2009	28°♌03'	May 1 2012	29°♌07'
Jun 1 2009	28°♌04'	Jun 1 2012	29°♌07'
Jul 1 2009	28°♌14'	Jul 1 2012	29°♌17'
Aug 1 2009	28°♌31'	Aug 1 2012	29°♌34'
Sep 1 2009	28°♌51'	Sep 1 2012	29°♌54'
Oct 1 2009	29°♌09'	Oct 1 2012	00°♍12'
Nov 1 2009	29°♌22'	Nov 1 2012	00°♍25'
Dec 1 2009	29°♌26'	Dec 1 2012	00°♍29'
Jan 1 2010	29°♌19'	Jan 1 2013	00°♍23'
Feb 1 2010	29°♌05'	Feb 1 2013	00°♍08'
Mar 1 2010	28°♌49'	Mar 1 2013	29°♌52'
Apr 1 2010	28°♌33'	Apr 1 2013	29°♌37'
May 1 2010	28°♌25'	May 1 2013	29°♌28'
Jun 1 2010	28°♌25'	Jun 1 2013	29°♌28'
Jul 1 2010	28°♌34'	Jul 1 2013	29°♌37'
Aug 1 2010	28°♌52'	Aug 1 2013	29°♌54'
Sep 1 2010	29°♌12'	Sep 1 2013	00°♍14'
Oct 1 2010	29°♌30'	Oct 1 2013	00°♍33'
Nov 1 2010	29°♌43'	Nov 1 2013	00°♍46'
Dec 1 2010	29°♌47'	Dec 1 2013	00°♍50'

JAN	1	2014	00° ♍ 44'
FEB	1	2014	00° ♍ 30'
MAR	1	2014	00° ♍ 14'
APR	1	2014	29° ♌ 58'
MAY	1	2014	29° ♌ 49'
JUN	1	2014	29° ♌ 49'
JUL	1	2014	29° ♌ 58'
AUG	1	2014	00° ♍ 15'
SEP	1	2014	00° ♍ 35'
OCT	1	2014	00° ♍ 53'
NOV	1	2014	01° ♍ 07'
DEC	1	2014	01° ♍ 11'
JAN	1	2015	01° ♍ 05'
FEB	1	2015	00° ♍ 51'
MAR	1	2015	00° ♍ 35'
APR	1	2015	00° ♍ 19'
MAY	1	2015	00° ♍ 10'
JUN	1	2015	00° ♍ 10'
JUL	1	2015	00° ♍ 18'
AUG	1	2015	00° ♍ 35'
SEP	1	2015	00° ♍ 55'
OCT	1	2015	01° ♍ 14'
NOV	1	2015	01° ♍ 27'
DEC	1	2015	01° ♍ 31'